REVISE BTEC NATIONAL
Applied Science

REVISION WORKBOOK

Series Consultant: Harry Smith

Authors: Cliff Curtis, Ann Fullick, Karlee Lees, Chris Meunier and Carol Usher

---

A note from the publisher

While the publishers have made every attempt to ensure that advice on the qualification and its assessment is accurate, the official specification and associated assessment guidance materials are the only authoritative source of information and should always be referred to for definitive guidance.

This qualification is reviewed on a regular basis and may be updated in the future. Any such updates that affect the content of this Revision Workbook will be outlined at www.pearsonfe.co.uk/BTECchanges.

For the full range of Pearson revision titles across KS2, KS3, GCSE, Functional Skills, AS/A Level and BTEC visit:
www.pearsonschools.co.uk/revise

Published by Pearson Education Limited, 80 Strand, London, WC2R 0RL.

www.pearsonschoolsandfecolleges.co.uk

Copies of official specifications for all Pearson qualifications may be found on the website: qualifications.pearson.com

Text and illustrations © Pearson Education Ltd 2018
Typeset and illustrated by Kamae Design
Produced by Out of House Publishing
Cover illustration by Miriam Sturdee

The rights of Cliff Curtis, Ann Fullick, Karlee Lees, Chris Meunier and Carol Usher to be identified as authors of this work have been asserted by them in accordance with the Copyright, Designs and Patents Act 1988.

First published 2018

21 20 19 18

10 9 8 7 6 5 4 3 2 1

British Library Cataloguing in Publication Data
A catalogue record for this book is available from the British Library

ISBN 978 1 292 15003 1

Copyright notice
All rights reserved. No part of this publication may be reproduced in any form or by any means (including photocopying or storing it in any medium by electronic means and whether or not transiently or incidentally to some other use of this publication) without the written permission of the copyright owner, except in accordance with the provisions of the Copyright, Designs and Patents Act 1988 or under the terms of a licence issued by the Copyright Licensing Agency, Barnards Inn, 86 Fetter Lane, London EC4A 1EN (www.cla.co.uk). Applications for the copyright owner's written permission should be addressed to the publisher.

Printed in Slovakia by Neografia

Acknowledgements
The authors and publisher would like to thank Chris Short for his assistance in the production of this book. We are grateful to the following for permission to reproduce copyright material:

(Key: b-bottom; t-top)

Photos
Alamy Stock Photo: Benedicte Desrus 111, Jack Picone 110, Jake Lyell 144, Mark Boulton 145, Michele Burgess 113, Phanie 112t, Stocktrek Images 7; **Anthony Short:** 149; **NASA:** 150; **Science Photo Library Ltd:** Biophoto Associates 21, Dr Keith Wheeler 4, GIPhotostock 37; **Shutterstock.com:** Michael Pettigrew 112b

Text
Page 117: reprinted by permission from Macmillan Publishers Ltd: *Nature Genetics;* Genetic architecture of artemisinin-resistant *Plasmodium falciparum*, copyright Jan 19, 2015; pages 112–113, 119: reproduced with permission from Mirrorpix; pages 144–146: © Telegraph Media Group Limited 2015; pages 150–152: reproduced with permission from Power-technology.com and Global Data.

All other images © Pearson Education Ltd

Every effort has been made to contact copyright holders of material reproduced in this book. Any omissions will be rectified in subsequent printings if notice is given to the publishers.

Notes from the publisher
1. While the publishers have made every attempt to ensure that advice on the qualification and its assessment is accurate, the official specification and associated assessment guidance materials are the only authoritative source of information and should always be referred to for definitive guidance.

    Pearson examiners have not contributed to any sections in this resource relevant to examination papers for which they have responsibility.

2. Pearson has robust editorial processes, including answer and fact checks, to ensure the accuracy of the content in this publication, and every effort is made to ensure this publication is free of errors. We are, however, only human, and occasionally errors do occur. Pearson is not liable for any misunderstandings that arise as a result of errors in this publication, but it is our priority to ensure that the content is accurate. If you spot an error, please do contact us at resourcescorrections@pearson.com so we can make sure it is corrected.

Websites
Pearson Education Limited is not responsible for the content of any external internet sites. It is essential for tutors to preview each website before using it in class so as to ensure that the URL is still accurate, relevant and appropriate. We suggest that tutors bookmark useful websites and consider enabling students to access them through the school/college intranet.

Introduction

This Workbook has been designed to help you revise the skills you may need for the externally assessed units of your course. Remember that you won't necessarily be studying all the units included here – it will depend on the qualification you are taking.

BTEC National Qualification	Externally assessed units
Certificate	Unit 1: Principles and Applications of Science I
For both: Extended Certificate Foundation Diploma	Unit 1: Principles and Applications of Science I Unit 3: Science Investigation Skills
Diploma	Unit 1: Principles and Applications of Science I Unit 3: Science Investigation Skills Unit 5: Principles and Applications of Science II
Extended Diploma	Unit 1: Principles and Applications of Science I Unit 3: Science Investigation Skills Unit 5: Principles and Applications of Science II Unit 7: Contemporary Issues in Science

Your Workbook

Each unit in this Workbook contains either one or two sets of revision questions or revision tasks to help you **revise the skills** you may need in your assessment. The selected content, outcomes, questions and answers used in each unit are provided to help you to revise content and ways of applying your skills. Ask your tutor or check the Pearson website for the most up-to-date **Sample Assessment Material** and **Mark Schemes** to get an indication of the structure of your actual assessment and what this requires of you. The detail of the actual assessment may change, so always make sure you are up to date.

This Workbook will often include one or more useful features that explain or break down longer questions or tasks. Remember that these features won't appear in your actual assessment.

> Grey boxes like this contain **hints and tips** about ways that you might complete a task, interpret a brief, understand a concept or structure your responses.

 This icon will appear next to **an example partial answer** to a revision question or task. You should read the partial answer carefully, then complete it in your own words.

> This is a **revision activity**. It will help you understand some of the skills needed to complete the revision task or question.

> 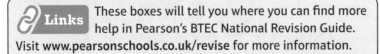 These boxes will tell you where you can find more help in Pearson's BTEC National Revision Guide.
> Visit www.pearsonschools.co.uk/revise for more information.

There is often space on the pages for you to write in. However, if you are carrying out research and making ongoing notes, you may want to use separate paper. Similarly, some units will be assessed through submission of digital files, or on screen, rather than on paper. Ask your tutor or check the Pearson website for the most up-to-date details.

Contents

Unit 1: Principles and Applications of Applied Science I

1   Your exam
2   Revision test 1
20   Revision test 2

Unit 3: Science Investigation Skills

39   Your set task
40   Revision task 1
52   Revision task 2

Unit 5: Principles and Applications of Applied Science II

64   Your exam
65   Revision test 1
86   Revision test 2

Unit 7: Contemporary Issues in Science

108   Your set task
109   Revision task 1
144   Revision task 2

163   Periodic table
164   Unit 1 formula sheet
165   Unit 5 formula sheet
166   Answers

A small bit of small print

Pearson publishes Sample Assessment Material and the Specification on its website. This is the official content and this book should be used in conjunction with it. The questions and revision tasks in this book have been written to help you practise the knowledge and skills you will require for your assessment. Remember: the real assessment may not look like this.

Unit 1: Principles and Applications of Applied Science I

Your exam

Unit 1 will be assessed through an exam, which will be set by Pearson. You will need to use your understanding of core science concepts to respond to questions that require short and long answers.

Your Revision Workbook

> This Workbook is designed to **revise skills** that might be needed in your exam. The details of the actual exam may change so always make sure you are up to date. Ask your tutor or check the **Pearson website** for the most up-to-date **Sample Assessment Material** to get an indication of the structure of your exam and what this requires of you.

To support your revision, this Workbook contains revision questions to help you revise the skills that might be needed in your exam. The revision questions are divided into three sections:
- Section A: Biology (Structures and functions of cells and tissues)
- Section B: Chemistry (Periodicity and properties of elements)
- Section C: Physics (Waves in communication).

 The periodic table of elements is on page 162 and the formulae sheet can be found on page 163.

Each of the sections will contain a range of different question types, including multiple choice, short answer, calculations, drawing questions and open response. You should make sure you understand what each different command word is asking you to do in the question.

 To help you revise skills that might be needed in your Unit 1 exam this Workbook contains two sets of revision questions starting on pages 2 and 20. The first set contains lots of help with hints and guided answers, while the second set is less guided to give you more practise at answering questions. See the introduction on page iii for more information on features included to help you revise.

- Read each question carefully before you start to answer it.
- Check your answers if you have time at the end.

Revision test 1

Section A: Structures and functions of cells and tissues

Answer ALL questions. Write your answers in the spaces provided.

1 The graph shows a recording of an action potential.

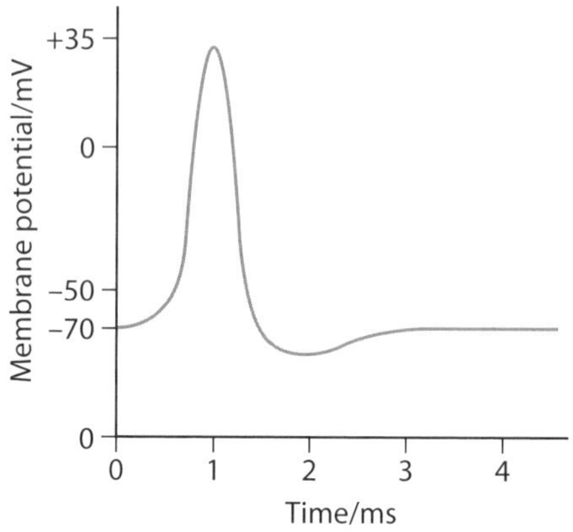

(a) What state is the membrane potential at 1 ms on the graph? **1 mark**

☐ A depolarised

☐ B hyperpolarised

☐ C polarised

☐ D repolarised.

(b) Sodium channels open to allow an increased flow of sodium ions into the neuron.

Use the graph to estimate the time when this happens. **1 mark**

....

> Look for the time when the membrane potential starts to increase for depolarisation. Also you are looking for a precise time not a range.

(c) State the time when hyperpolarisation is at its greatest. **1 mark**

....

> In action potentials remember depolarised is when membrane potential is positive and polarised is when the membrane potential is negative. 'Hyper' means over or above, so this will be the membrane potential when it is at the **most** negative membrane potential.

(d) At what time do the potassium channels open? **1 mark**

☐ A 1 ms

☐ B 2 ms

☐ C 3 ms

☐ D 0.5 ms

> Potassium ions diffuse out of the cell and take positive charge with them, making the cell potential more negative. When do the cells start becoming more negative?

Guided (e) Explain how the normal balance of sodium and potassium ions is regained. 2 marks

The Na⁺/K⁺ ATP pumps use energy to rapidly transport ....

....

....

....

Links You can revise action potential on page 11 of the Revision Guide.

Total for Question 1 = 6 marks

Unit 1 Guided

2 The photograph shows a transverse section of a pondweed stem through a light microscope.

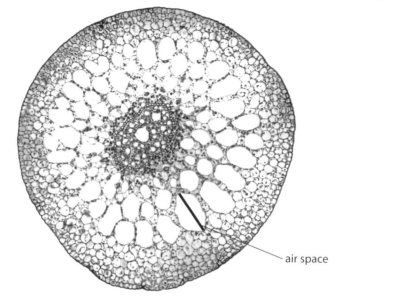

air space

(a) The total magnification of the image is ×50. The eyepiece lens has a magnification of ×10.

Calculate the magnification of the objective lens. **2 marks**

total magnification = eyepiece lens × the objective lens

Magnification = ...

Guided

(b) Calculate the actual length of the air space marked on the photograph. **2 marks**

$$\text{magnification} = \frac{\text{size of image}}{\text{size of real object}}$$

$$\text{size of real object} = \frac{\text{size of image}}{\text{magnification}}$$

You need to measure the actual length of the air space first. Be careful of the units.

$$\text{length of air space} = \frac{}{500} = ... \mu m$$

(c) All aquatic plants have a similar stem structure, with a cortex of parenchyma cells with large air spaces. Suggest an explanation of why the stems have a structure like this. **3 marks**

....

....

....

Consider what an aquatic plant needs to do in order to survive.

....

....

....

....

Total for Question 2 = 7 marks

3 Plants are complex multi-cellular organisms and have many different specialised cell types.

(a) Give a definition for specialised cells. **2 marks**

....
....
....
....

(b) Below are diagrams of two specialised plant cells A and B.

(i) State the names and functions of cell A and cell B. **2 marks**

Cell A is a .... it carries out ....

Cell B is a .... it takes up ....

(ii) Compare cells A and B.

In your answer explain the differences between the two cells. **4 marks**

> You should use the number of marks as a guide for how many points to write. If you are asked to compare two things, one of those points must be a similarity or a difference.

....
....
....
....
....
....

Links You can revise specialised plant cells on page 5 of the Revision Guide.

Total for Question 3 = 8 marks

Unit 1
Guided

4 Myoglobin is a protein found in muscle cells, which binds on to oxygen.
There are two different types of muscle cells. They can be fast twitch or slow twitch.

Fast twitch muscle cells rely mainly on anaerobic respiration for energy. Anaerobic means without oxygen.

(a) Choose the statement below, which is **true**. *(1 mark)*

☐ **A** Slow twitch muscle cells have less myoglobin than fast twitch muscle cells.

☐ **B** Slow twitch muscle cells have more myoglobin than fast twitch muscle cells.

☐ **C** Slow twitch muscle cells have no myoglobin.

☐ **D** Slow twitch muscle cells have the same quantity of myoglobin as fast twitch muscle cells.

(b) Skeletal muscles have a bundle-within-bundles structure. They are covered in a tough layer of connective tissue, within this are many bundles, each of which contains 10 to 100 or more muscle fibres or cells.

Muscle cells also have a bundle-within-bundle structure. Describe the structure of a muscle cell, in terms of the muscle's main function of contraction. *(2 marks)*

It's important to read the question and note down what the question tells you. For this question, your answer should focus on features specific to a muscle cell, not general features of any cell such as a nucleus or cell membrane.

....

....

....

....

Total for Question 4 = 3 marks

Unit 1
Guided

5 Sickle-cell disease is an inherited condition, where the haemoglobin is abnormal.
This causes the red blood cells to be inelastic.
They only last for about 10–20 days compared to 3–4 months normally.
Below is an image showing sickle cells next to some normal red blood cells.

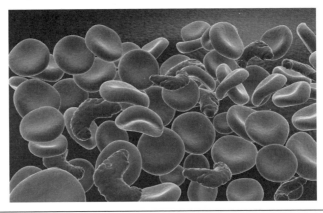

Deduce the problems these abnormal red blood cells could cause people who suffer from sickle-cell disease.

Your answer should explain the effects of the inelastic red blood cells and their short life span.

6 marks

> Remember that normal red blood cells squeeze through capillaries and that they carry oxygen to cells for respiration.

....
....
....
....
....
....
....
....
....
....
....

Links You can revise red blood cells on page 6 of the Revision Guide.

Total for Question 5 = 6 marks

END OF SECTION
TOTAL FOR SECTION A = 30 MARKS

Section B: Periodicity and properties of elements

Answer ALL questions. Write your answers in the spaces provided.

1   Bromine is a fuming red–brown liquid at room temperature. It is a mixture of two isotopes, ^{79}Br and ^{81}Br.

(a) Which of the following is the number of neutrons in the isotope ^{79}Br? **1 mark**

☐ **A** 42

☐ **B** 44

☐ **C** 46

☐ **D** 48

(b) Identify **two** correct statements about the isotopes. **2 marks**

☐ **A** ^{79}Br and ^{81}Br have the same number of atoms.

☐ **B** ^{79}Br and ^{81}Br have the same number of electrons.

☐ **C** ^{79}Br and ^{81}Br have the same number of ions.

☐ **D** ^{79}Br and ^{81}Br have the same number of neutrons.

☐ **E** ^{79}Br and ^{81}Br have the same number of protons.

> Isotopes are atoms with the same atomic number but different mass numbers.

Guided

(c) The electronic configuration of a bromine atom can be written in terms of sub-shells.

(i) Complete the electronic configuration of a bromine atom. **1 mark**

> Remember that bromine is in group 7 of the periodic table.

$1s^2\ 2s^2\ 2p^6\ 3s^2\ 3p^6$ ....

(ii) State why bromine is classified as a p-block element. **1 mark**

....

....

> Bromine fluoride is a polar molecule.

(d) Describe the charge distribution in a polar molecule. **1 mark**

....

....

Unit 1
Guided

> Bromine forms three compounds with phosphorus.
> The compounds have the molecular formulae PBr$_3$, PBr$_5$ and P$_2$Br$_4$

(e) (i) Give the meaning of molecular formula. **2 marks**

....

....

....

....

> PBr$_3$ can be prepared by heating phosphorus (P$_4$) in bromine vapour.

Guided

(ii) Write a chemical equation for this reaction. State symbols are not required. **1 mark**

P$_4$ + ... Br$_2$ → .... PBr$_3$

(iii) One of the three phosphorus bromides has the following percentage composition by mass:

P = 16.2%   Br = 83.8%

Calculate the empirical formula of this bromide. **2 marks**

> The percentages are the same as the mass, in grams, in 100 g of the compound. The calculation should be carried out in three stages:
> 1. Calculate the amount, in moles, of each element in 100 g (divide the mass by the relative atomic mass, Ar).
> 2. Calculate the ratio of moles, by dividing both by the smaller value.
> 3. Calculate, if necessary, the simplest whole number ratio of moles of each element.
> Now write the empirical formula using this ratio.

....

....

....

....

empirical formula ....

(iv) Determine the identity of the bromide. **1 mark**

....

> **Links** You can revise moles on page 22 of the Revision Guide.

Total for Question 1 = 12 marks

2 The bar chart shows the first ionisation energies of the elements sodium to potassium.

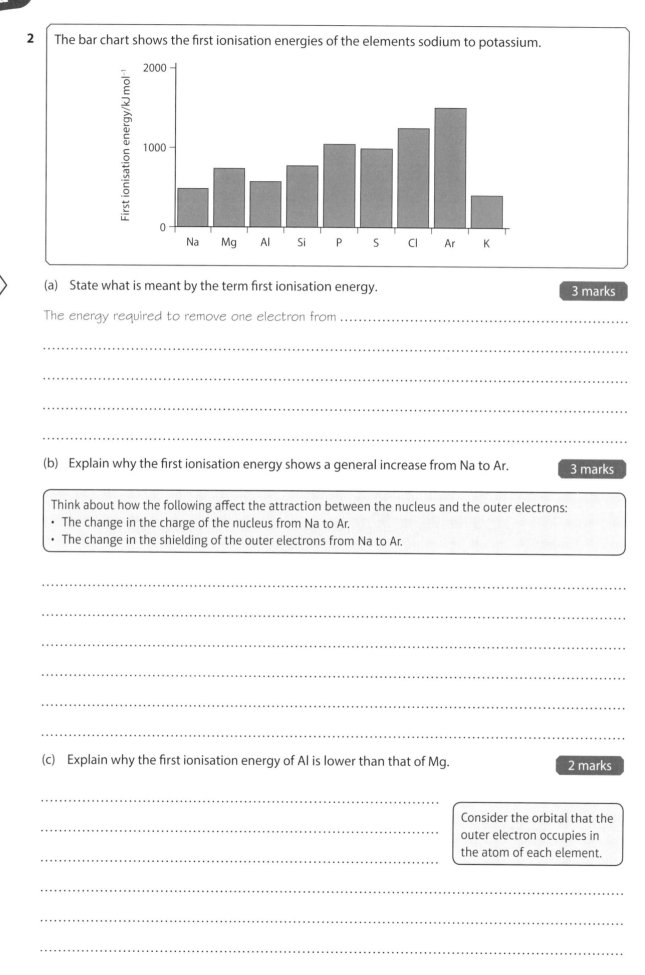

(a) State what is meant by the term first ionisation energy. **3 marks**

The energy required to remove one electron from ....

(b) Explain why the first ionisation energy shows a general increase from Na to Ar. **3 marks**

> Think about how the following affect the attraction between the nucleus and the outer electrons:
> • The change in the charge of the nucleus from Na to Ar.
> • The change in the shielding of the outer electrons from Na to Ar.

(c) Explain why the first ionisation energy of Al is lower than that of Mg. **2 marks**

> Consider the orbital that the outer electron occupies in the atom of each element.

(d) Explain why the first ionisation energy of K is much lower than that of Ar, even though it has a higher nuclear charge.

4 marks

Include the following in your answer:
- In which quantum shell each outer electron can be found.
- How much shielding the outer electron experiences in each atom.

....
....
....
....
....
....
....
....

Total for Question 2 = 12 marks

Unit 1 Guided

Guided 3 Chlorine, bromine and iodine are known as halogens.

Discuss how van der Waals forces are responsible for the trend in boiling temperature of these halogens.

6 marks

To 'discuss', you need to show a clear understanding of the topic, consider all aspects of the topic, make connections between different aspects and discuss the extent or importance of features.

Consider the following points:
- How an instantaneous dipole is created in a molecule.
- How this instantaneous dipole affects a neighbouring molecule.
- How the number of electrons in a molecule affects the size of the instantaneous dipole and hence the size of the attraction between molecules.
- How the number of electrons in the molecule changes from chlorine to bromine to iodine.
- How the size of the attraction between the molecules affects the amount of energy required to separate the molecules to form a gas.
- How the amount of energy required to separate the molecules affects the boiling temperature.

....
....
....
....
....
....
....
....
....
....
....
....

Links You can revise intermolecular forces on page 20 of the Revision Guide.

Total for Question 3 = 6 marks

END OF SECTION TOTAL FOR SECTION B = 30 MARKS

Section C: Waves in communication

> Answer ALL questions. Write your answers in the spaces provided.

1 Wi-Fi and Bluetooth® are used to transmit data between instruments and computers, e.g. in a hospital ward or operating theatre. By using different protocols they can successfully share the same frequency range.

(a) In which of these regions of the electromagnetic spectrum do Wi-Fi and Bluetooth® data communications operate? **1 mark**

- [ ] **A** 1 MHz
- [ ] **B** 2.5 GHz
- [ ] **C** 30 GHz
- [ ] **D** 1 THz

(b) A TV remote control operates on a wavelength of 940 nm. Select the region of the electromagnetic spectrum in which this wavelength falls. **1 mark**

- [ ] **A** microwave
- [ ] **B** visible light
- [ ] **C** infrared
- [ ] **D** ultraviolet

Links See page 40 of the Revision Guide for reminders of the regions and the frequency bands of the electromagnetic spectrum.

> **Guided**
>
> A mobile phone handset is 1 km from the nearest cellular mast and 3 km from the mast in an adjacent cell.

(c) Explain how different frequency bands are used by mobile phone networks to achieve interference-free communications. **3 marks**

Each mobile phone operator is allocated a separate set of

frequency bands by the ....

....

For three marks you need to make three separate points.

The upload signal from a mobile handset is broadcast on a slightly different frequency from the

....

The cell masts in adjacent cells ....

....

Unit 1
Guided

Guided

(d) The upload signal broadcast from the phone handset is detected at both the masts shown in the diagram.

Calculate how many times stronger the detected signal will be at the nearest mast, compared with that at the mast in the next-door cell.

2 marks

The broadcast is in all directions, so its intensity obeys the inverse square law: $I = k/r^2$

So $I_1 / I_2 = (r_2 / r_1)^2 = (\ldots\ldots\ldots\ldots\text{km} / 1\,\text{km})^2$

$= \ldots\ldots\ldots\ldots$ times stronger.

Quote the law or formula you are using, and show your calculation working. There is a mark for each.

Total for Question 1 = 7 marks

2 A clarinet has a mouthpiece with a single reed. That end of the clarinet behaves like a closed-ended pipe. The other end of the instrument is open to the air.

When air is blown over the reed in the clarinet mouthpiece, a stationary wave is set up in the instrument and a sound is produced.

(a) A sound wave is produced by compressions and rarefactions of the material through which it travels. Name this type of wave. **1 mark**

....

The diagram shows how the amplitude of air vibrations varies down the length of such a pipe

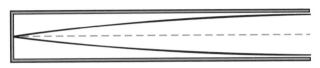

(b) Label **one** node and **one** antinode on the diagram. **2 marks**

Node means a point of no vibration.

(c) The pipe is 0.75 m long. Calculate the wavelength for the next higher harmonic above the fundamental mode. **2 marks**

The next harmonic has two nodes and still has an antinode at the open end.

So it fits ¾ of a wavelength into the pipe. So, ¾ λ = ...

Therefore λ = ...

Links You can revise musical instruments on page 36 of the Revision Guide.

Total for Question 2 = 5 marks

Unit 1 Guided

3 All four strings on a violin are the same length, but they have different mass per unit length values and can be separately tensioned to tune them. In the head of a violin, each string is wound round a peg, which can be turned to adjust the tension in that string.

Guided

(a) (i) Calculate how many times larger the tension force must be to double the frequency of the note.

2 marks

> Look on the formula sheet (page 163) to find equations: 'wave speed' and 'speed of a transverse wave on a string'.

$f = v/\lambda$

$v = (T/\mu)^{1/2}$

It is the same string throughout, so μ has a constant value.

Therefore, $v \propto T^{1/2}$

So, if f_1 is the original frequency, and f_2 is the frequency of the note one octave higher:

$f_2/f_1 = 2 = v_2/v_1 = (T_2/T_1)^{1/2}$

Because the length of the string does not change, λ is constant, so can be disregarded in this equation.

So, $(T_2/T_1) = 2^2 = $ .... times larger.

(ii) Give **two** differences between the types of wave produced in the string when the violin bow is drawn across it and the sound waves that radiate out from it through the air.

2 marks

> Waves can be progressive or stationary. They can also be either transverse or longitudinal.

....

....

....

....

....

(iii) Interference patterns occur when waves are coherent. State what is meant when two waves are coherent.

1 mark

....

The diagram shows a spectrum made when light, produced by passing an electric discharge through a mercury vapour sample, is passed through a diffraction grating. This spectrum is a series of separate lines of specific colours.

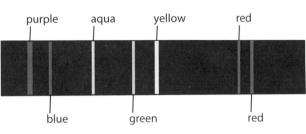

(b) Describe how emission spectra such as the one shown above can be used to identify elements in samples of unknown composition.

2 marks

....
....
....
....

Links You can revise interference and diffraction gratings on pages 33–34 of the Revision Guide.

Total for Question 3 = 7 marks

4 A table decoration can be created using optical fibres made from Perspex®. A light source in the base of the decoration feeds light into the cut ends of a bunch of fibres. The fibres are allowed to spread out at their other cut ends, where bright dots of light appear. The fibres appear to be unlit along their length apart from those bright spots.

(a) Perspex® has a refractive index of 1.48.

Calculate the critical angle for Perspex®. **2 marks**

Refractive index, n = 1/sin C. So sin C = 1/n = ...

Therefore the critical angle, C = sin⁻¹ ....

= ....

Find the formula connecting critical angle with refractive index on the formula sheet and rearrange it.

(b) Perspex® is a transparent material, but light emerges only from the ends of the fibres.

Draw the path taken by one of the light rays passing through a fibre on the diagram. **2 marks**

(c) Explain why no light is seen coming out from the sides of the fibres. **1 mark**

Most of the light that enters the fibre will strike the outside surface at an angle

.... than the critical angle and so will undergo

....

....

Unit 1
Guided

> Optical fibres make important contributions to improved healthcare.
>
> One use of them is in endoscopy. This technique, widely used in hospitals, allows many types of medical investigation and operation to be performed with only minimal cutting of body tissue.
>
> To transmit and receive patient data, hospitals, GPs and other health professionals also increasingly rely on broadband internet connections that use fibre optics.

(d) Compare the uses made of optical fibres in endoscopy and in broadband internet connections.

In your answer you should refer to frequencies, analogue and digital data types, the types of fibre used and the quality of the data communicated. **6 marks**

In a medical endoscope, the optical fibres can carry light with a full range of visible frequencies.

One bundle of fibres carries ....

Another bundle of fibres ....

....

....

....

By contrast, fibres in a broadband network ....

That means ....

> To demonstrate your understanding of the topic, it helps to explain the term 'digital' by describing how numbers are encoded by flashes of light.

Each separate frequency band of IR radiation can carry another separate set of digital information. The fibres used in an endoscope do not need to carry the signal more than about a metre and just need to deliver a lot of light efficiently. But in broadband optical fibre cables ....

....

....

....

....

....

....

....

Links You can revise fibre optics and types of signal on pages 37–39 of the Revision Guide.

Total for Question 4 = 11 marks

END OF PAPER

TOTAL FOR SECTION C = 30 MARKS
TOTAL FOR PAPER = 90 MARKS

Revision test 2

Section A: Structures and functions of cells and tissues

> Answer ALL questions. Write your answers in the spaces provided.

1 Below is a diagram of a plant cell.

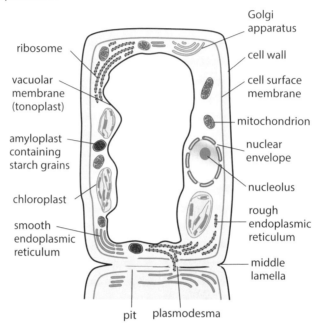

(a) Plant cells have some different features to animal cells.

Compare a plant cell with an animal cell.

3 marks

> Compare means you need to state any similarities or differences between a plant cell and an animal cell. You should use the number of marks as a guide for how many points to write.

....
....
....
....
....
....
....

Links You can revise plant and animal cells on pages 4–6 of the Revision Guide.

(b) The image below shows an electron micrograph of a plant cell.

Calculate the diameter X, marked on a chloroplast in micrometres.

3 marks

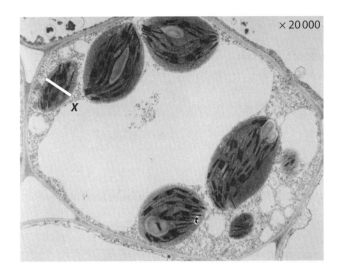

magnification = size of image/size of real object

There are 1000 micrometres in a millimetre.

Diameter X of chloroplast = ... μm

 You can revise epithelial cells on page 7 of the Revision Guide.

Total for Question 1 = 6 marks

Unit 1

2 The diagram below shows squamous epithelium, which is found in the human lung.

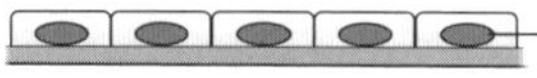

apical side (top)

● ● ● ● ● —— nuclei

basal side (base)

(a) What part of the human lung contains mainly squamous epithelium? **1 mark**

☐ **A** alveoli

☐ **B** bronchi

☐ **C** trachea

☐ **D** bronchioles.

(b) Explain how the structure and function of squamous epithelium in the lung are important for gas exchange. **4 marks**

....

....

....

....

> In this question you need to state two features of squamous epithelium which help gas exchange, and the reasons why.

....

....

....

....

> Cigarette smoke can cause emphysema. This is a chronic obstructive pulmonary disorder (COPD) that causes shortness of breath.

(c) Explain why people with emphysema suffer from shortness of breath. **3 marks**

....

....

> Cigarette smoke kills alveoli cells.

....

....

....

....

Links You can revise alveoli on page 7 of the Revision Guide.

Total for Question 2 = 8 marks

3 The diagram shows a prokaryotic cell.

(a) Complete the labelling.   [3 marks]

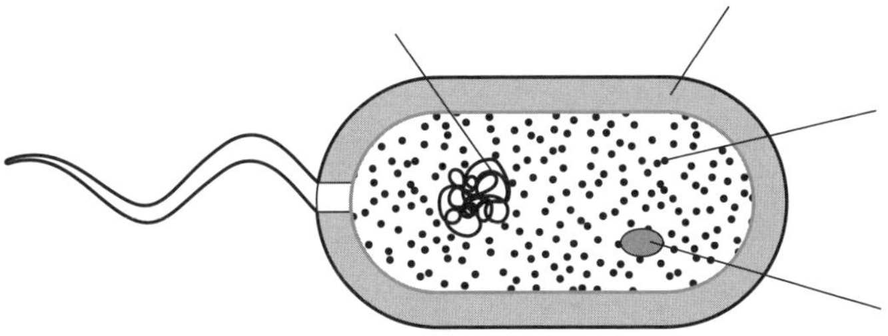

> Remember that prokaryotes do not have a membrane-bound nucleus.

(b) Bacteria can be Gram-positive or Gram-negative.

Describe how a scientist can distinguish between them.   [2 marks]

> This question only asks how you can test whether a bacterium is Gram-negative or Gram-positive and does not ask you to explain why this happens.

....

....

....

....

(c) Explain why Gram-negative bacteria do not stain.   [1 mark]

....

....

....

> This only asks about Gram-negative bacteria, so you do not need to mention Gram-positive bacteria.

Links You can revise bacteria on page 3 of the Revision Guide.

Total for Question 3 = 6 marks

4 Nerve impulses are transmitted along the axon of a neuron.

(a) The diagram shows the structure of a motor neuron.

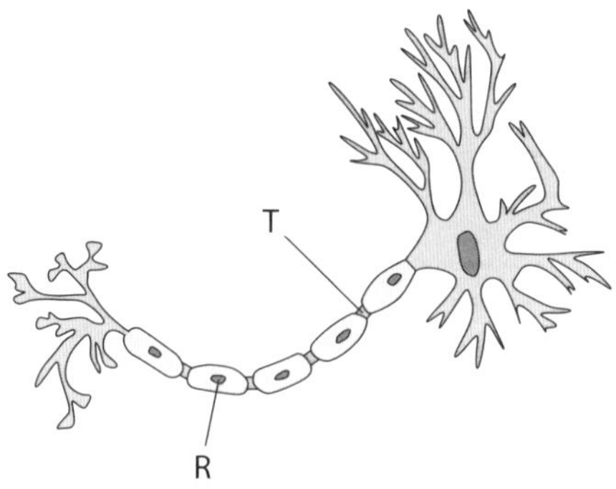

Synapses occur between cells.
A motor neuron is a cell.

   (i) Name the part of the neuron labelled T. *1 mark*

      ☐ **A** dendrite

      ☐ **B** node of Ranvier

      ☐ **C** Schwann cell

      ☐ **D** synapse.

   (ii) Name the part of the neuron labelled R. *1 mark*

      ☐ **A** synapse

      ☐ **B** nucleus of cell body

      ☐ **C** dendrite

      ☐ **D** nucleus of Schwann cell.

(b) Explain why most nerve cells have a myelin sheath. *2 marks*

> Think of a myelin sheath as the plastic surrounding an electrical wire.

(c) The signal is transmitted between neurons across a synapse.

Explain the sequence of events that occur at the synapse after a neurotransmitter has been released.

6 marks

Aim to give a detailed explanation. Make sure your answer is clear and logically structured and that your points are linked.

Links You can revise synapses on page 13 of the Revision Guide.

You should include what a synapse is.

Total for Question 4 = 10 marks

END OF SECTION TOTAL FOR SECTION A = 30 MARKS

Unit 1

Section B: Periodicity and properties of elements

Answer ALL questions. Write your answers in the spaces provided.

The periodic table can be used to predict the chemical properties of elements. Part of the periodic table is shown below

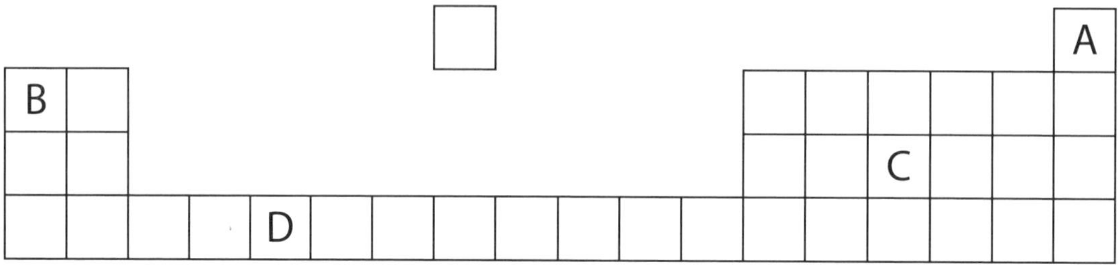

[Periodic table diagram with letters A (top right), B (top left, group 1 row 2), C (right side), D (bottom, group around 3)]

The letters A, B, C and D represent four different elements.

1 (a) (i) Which element, A, B, C or D, has atoms with a 3p sub-shell containing three electrons?   *1 mark*

☐ A
☐ B
☐ C
☐ D

(ii) Which element, A, B, C or D, has the highest first ionisation energy?   *1 mark*

☐ A
☐ B
☐ C
☐ D

(b) Describe how an ion is formed from an element found in group 2 of the periodic table.   *2 marks*

....
....
....
....

🔗 **Links** You can revise the periodic table on page 25 of the Revision Guide.

Total for Question 1 = 4 marks

Unit 1

2 The table shows the boiling temperatures of the elements sodium to chlorine in period 3 of the periodic table.

Element	Na	Mg	Al	Si	P	S	Cl
Melting temperature (°C)	98	650	660	1410	44	113	−101
Boiling temperature (°C)	890	1110	2470	2360	280	445	−35

(a) Which of the following elements has metallic bonding? **1 mark**

☐ **A** Cl

☐ **B** Na

☐ **C** P

☐ **D** Si

(b) (i) Draw a labelled diagram to show the arrangement of the particles involved in the bonding of a metal. **2 marks**

(ii) Describe what is meant by the term metallic bonding. **2 marks**

....

....

....

....

(c) Explain the difference in melting temperature between magnesium and sodium. **4 marks**

> Think about the charge on the sodium and magnesium ions and the number of delocalised electrons per atom.

....

....

....

....

....

....

Links You can revise metallic bonding on page 19 of the Revision Guide.

Total for Question 2 = 9 marks

3 The equation shows the reaction between barium and water.

$$Ba(s) + 2H_2O(l) \rightarrow Ba(OH)_2(aq) + H_2(g)$$

(a) Explain why this reaction is classified as a redox reaction. Use oxidation numbers in your answer. **2 marks**

Think about the word 'redox'.

....

....

....

....

(b) A sample of barium of mass 2.74 g is completely reacted with water to form 250 cm³ of aqueous barium hydroxide.

   (i) Calculate the amount of barium reacted.

   [Molar mass of Ba = 137 g mol⁻¹] **1 mark**

   Remember the equation relating moles and molar mass.

   amount = ... mol

   (ii) Calculate the volume, in cm³, of hydrogen that is produced. **2 marks**

   [Assume that 1 mol of hydrogen occupies 24 dm³ under the conditions used in the experiment.]

   volume = ... cm³

   (iii) Calculate the concentration of the aqueous barium hydroxide formed, in mol dm⁻³. **2 marks**

   concentration = ... mol dm⁻³

Links You can revise oxidation and reduction on page 29 of the Revision Guide.

Total for Question 3 = 7 marks

4 The properties of halogens mean that they have many industrial and domestic uses.

(a) Explain **one** reason why compounds of chlorine are added to public swimming pools. [2 marks]

> In 'explain' questions, you must show that you understand the topic and give reasons to support your answer.

....
....
....
....

(b) Bromine is also sometimes added to swimming pools.

Give **two** other uses of bromine compounds. [2 marks]

....
....
....
....
....
....

Unit 1

The reactivity of halogens can be seen from colour changes when experimenting using the chemicals below.
- aqueous sodium bromide
- aqueous sodium iodide
- aqueous bromine
- aqueous chlorine
- aqueous iodine.

(c) Explain how **three** experiments can be used to establish the order of reactivity of the halogens bromine, chlorine and iodine.

6 marks

In your explanation you should outline the reactants, products and observations that would indicate the order of reactivity. You must show that you understand why the observations happen and how they can be used for this purpose.

....
....
....
....
....
....
....
....
....
....
....
....

Links You can revise displacement reactions of halogens on page 29 of the Revision Guide.

Total for Question 4 = 10 marks

END OF SECTION
TOTAL FOR SECTION B = 30 MARKS

Section C: Waves in communication

Answer ALL questions. Write your answers in the spaces provided.

1 The diagram below illustrates how the analogue signal from a microphone might be sampled by an analogue to digital converter.

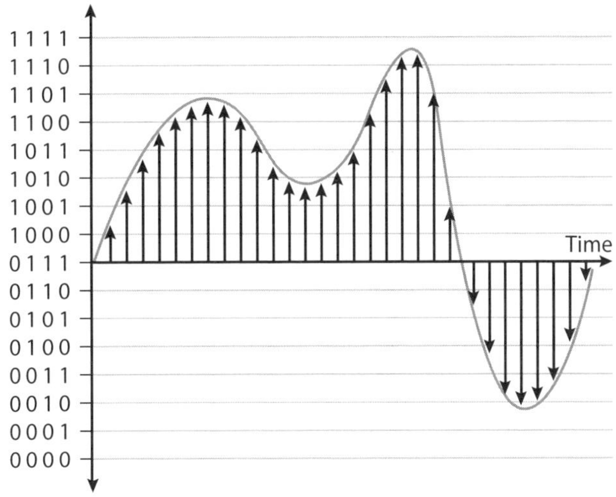

```
1 1 1 1
1 1 1 0
1 1 0 1
1 1 0 0
1 0 1 1
1 0 1 0
1 0 0 1
1 0 0 0
0 1 1 1
0 1 1 0
0 1 0 1
0 1 0 0
0 0 1 1
0 0 1 0
0 0 0 1
0 0 0 0
```

The output is a string of numbers represented by binary code bytes.

(a) (i) Explain the impact of sampling rate and sampling sensitivity on the quantity of data that has to be transmitted and also on the quality of the sound reproduction. **2 marks**

(ii) State **three** advantages of using digital signals in long-distance communication. **3 marks**

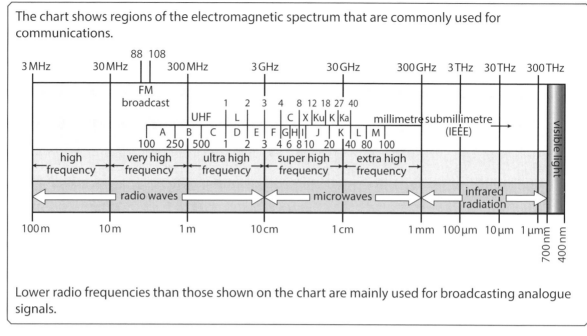

Lower radio frequencies than those shown on the chart are mainly used for broadcasting analogue signals.

(b) Explain why digital broadcasting requires the higher bandwidths that are possible with higher frequencies, and why infrared is not suitable for satellite communications. **3 marks**

> Compare the quantities of data involved in digital and in analogue representation of sound or images. What happens to signals as they travel through the atmosphere?

....
....
....
....
....
....

Links You can revise analogue to digital conversion on page 39 of the Revision Guide.

Total for Question 1 = 8 marks

2 The graph shows two sound waves from separate sources A and B, each detected by a microphone that is fixed at a particular point in space.

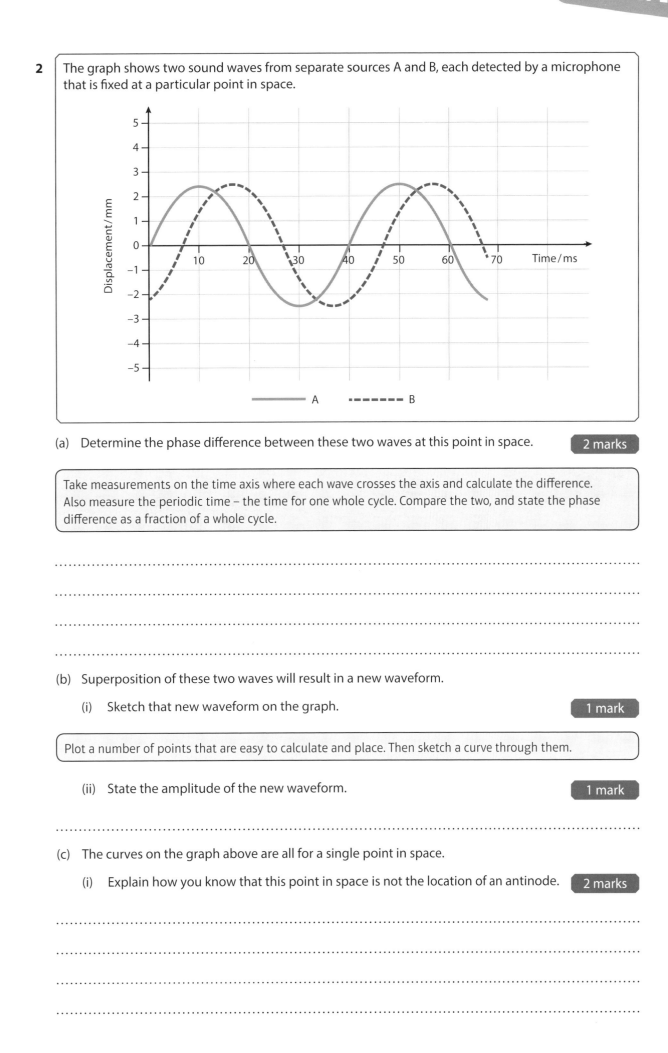

(a) Determine the phase difference between these two waves at this point in space. **2 marks**

> Take measurements on the time axis where each wave crosses the axis and calculate the difference. Also measure the periodic time – the time for one whole cycle. Compare the two, and state the phase difference as a fraction of a whole cycle.

....
....
....
....

(b) Superposition of these two waves will result in a new waveform.

　(i) Sketch that new waveform on the graph. **1 mark**

> Plot a number of points that are easy to calculate and place. Then sketch a curve through them.

　(ii) State the amplitude of the new waveform. **1 mark**

....

(c) The curves on the graph above are all for a single point in space.

　(i) Explain how you know that this point in space is not the location of an antinode. **2 marks**

....
....
....
....

(ii) State what the resultant amplitude would be at an antinode.

1 mark

....
....
....

What would the phase difference be at an antinode?

Links You can revise wave types on page 31, and superposition and interference on page 33 of the Revision Guide.

Total for Question 2 = 7 marks

3 Guitar strings need to have different mass per unit length to help them produce different notes. There are six strings on a guitar. The lowest and the highest are both tuned to the same musical note, 'E'. The 'top E' is two octaves higher than the 'bottom E' string, i.e. its frequency is four times higher.

(a) The tension on each of the strings is designed to be approximately the same.

The mass per unit length for the top E string is $1.0\,\text{g}\,\text{m}^{-1}$

Calculate what the mass per unit length of the bottom E string should be. **2 marks**

State the equation (from the formula sheet), rearrange it to obtain mass per unit length, then show all your numerical working.

(b) While the string is being plucked, if a finger is gently placed exactly halfway down a string and then quickly removed, the string plays a note one octave higher than its usual note.

  (i) Sketch the string to show how amplitude varies along its length in that higher mode. **1 mark**

  (ii) Explain why placing the finger on it has that effect. **2 marks**

  ....
  ....
  ....
  ....
  ....

  Think about nodes and antinodes.

(c) The frequency of the bottom E note on the guitar is 164.8 Hz. The speed of sound in air is 345 m s^{-1}. Calculate:

(i) the periodic time of the wave oscillations

1 mark

Remember Hz = s^{-1}

(ii) the wavelength of the sound waves in air for that note.

1 mark

Links You can revise the equations on page 32 of the Revision Guide.

Total for Question 3 = 7 marks

Unit 1

4 Lasers amplify light using a process called stimulated emission of radiation. The width of line in the spectrum produced by a laser is a measure of its coherence, and varies according to the type of laser source.

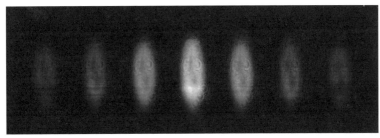

The photograph shows the diffraction pattern produced by laser light passing through a diffraction grating.

The table gives some examples taken from scientific papers of linewidth measurements for both pulsed and continuous wave types of laser.

Laser source	Wavelength (nm)	Frequency (THz)	Typical linewidth
High-power continuous wave laser (He–Ne single-transverse-mode)	633		1×10^6 Hz
Low-power continuous wave laser (cooled, stabilised, solid state)		563	1×10^3 Hz
High-gain high-power pulsed laser (rhodamine 6G dye)	590	508	5 to 10 nm (4 to 8×10^{12} Hz)
Low-power pulsed laser (multiple prism grating oscillator)	590	508	350×10^6 Hz

(a) Complete the table to show both wavelength and frequency values for each laser source. Show your working below.   **2 marks**

....

....

Start by quoting the equation for wave speed.

....

....

....

(b) State the meaning of the term 'coherence'.   **2 marks**

....

....

....

....

(c) Explain how diffraction gratings can be used to find the wavelength of a laser source. Use concepts of coherence, path difference, phase difference and superposition in your explanation.

4 marks

> Start by describing a diffraction grating and the process by which it produces coherent waves. Explain how path difference occurs and how this affects the phase difference of waves. Describe the condition for constructive interference.

....
....
....
....
....
....
....
....
....
....
....
....
....
....
....
....
....

> Check through your answer to make sure that you have demonstrated your understanding of the relationship between coherence, superposition and interference.

Links You can revise interference and diffraction gratings on pages 33–34 of the Revision Guide.

Total for Question 4 = 8 marks

END OF PAPER

TOTAL FOR SECTION C = 30 MARKS
TOTAL FOR PAPER = 90 MARKS

Unit 3: Science Investigation Skills

Your set task

Unit 3 will be assessed through a task, which will be set by Pearson. You will need to use your understanding of how to plan, record, process, analyse and evaluate scientific findings using primary and secondary data. You then answer questions based on practical work that has previously been carried out.

Your Revision Workbook

> This Workbook is designed to **revise skills** that might be needed in your assessed task. The details of the actual assessed task may change, so always make sure you are up to date. Ask your tutor or check the **Pearson website** for the most up-to-date **Sample Assessment Material** to get an indication of the structure of your assessed task and what this requires of you.

To support your revision, this Workbook contains two revision tasks to help you revise the skills that might be needed in your assessed task. The first of these tasks has more help and guidance, while the second allows you to apply the skills you have learnt.

In your assessment you may be asked to follow a method provided to carry out a practical investigation and record your results. You might later be expected to record your results and use these in answering questions. You are likely to need to be able to draw graphs and carry out calculations using practical data, as well as being able to write a plan for a given investigation and evaluate an investigation carried out by someone else.

The revision tasks in this Workbook have the practical results given for you so you do not need to carry out any practical work in order to use the Revision Workbook. However, you should make sure that you are clear on all the skills that you will need in your actual Unit 3 assessed task.

> **Links** To help you revise skills that might be needed in your Unit 3 set task this Workbook contains two revision tasks starting on pages 40 and 52. See the introduction on page iii for more information on features included to help you revise.

Revision task 1

Revision task brief

You are a research scientist working for a biotechnology company that carries out research on the industrial uses of enzymes. You have been asked to investigate the effect of pH on the enzyme amylase. Amylase is an enzyme which can break down starch and is often added to commercially available flour used in bread making. In this context amylase breaks down the starch in the flour into simpler sugars. Yeast then uses these sugars as respiratory substrates which produce carbon dioxide, causing the bread to rise during the fermentation process.

Amylase is an enzyme that will break down starch.

Iodine turns blue-black in the presence of starch and remains orange if there is no starch present.

Starch ⟶ Maltose

Blue-black in the presence of iodine   Orange in the presence of iodine

Practical investigation method

You must observe safe practice when carrying out the practical investigation. This method investigates the optimum pH for amylase to break down starch:

> **Links** For a reminder about the effect of pH on enzymes look at page 62 in the Revision Guide.

1. Label four test tubes pH 3, 4, 7 and 10 then add $4\,cm^3$ of the appropriate pH buffer to each (pH 3, 4, 7 and 10).

2. Add $4\,cm^3$ of amylase solution to each tube and place in a water bath at 37 °C for 5 minutes.

3. Add $4\,cm^3$ of 1% starch solution to each test tube and leave for 15 minutes.

4. Add four drops of each mixture to a separate trough on a spot plate.

5. Add one drop of iodine reagent to each.

6. Time how long it takes for the mixture to turn orange.

> So that you do not have to carry out any practical work in order to revise using this Workbook, we have provided some sample results for you. You should use these to answer the rest of the questions.

Investigation results

pH 3 – 470 seconds, 490 seconds, 450 seconds

pH 4 – 260 seconds, 230 seconds, 230 seconds

pH 7 – 60 seconds, 80 seconds, 70 seconds

pH 10 – 420 seconds, 450 seconds, 450 seconds.

Unit 3
Guided

You must complete ALL activities in Section 1 and Section 2.

Section 1

  1 (a) Record the experimental results in a suitable table, using the space provided.   **3 marks**

> You may need to bring results from your own practical investigation into your supervised assessment time. Check with your tutor or look at the latest Sample Assessment Material on the Pearson website for more information.
> To calculate the average, you add together your results for each pH and divide by the number of repeats. Always use the same number of decimal places for all numbers in the table.

pH	Time taken for iodine to turn orange (seconds)			
	1	2	3	Mean
3		490		
		80		
10				

Links For a reminder about drawing scientific tables look at page 76 in the Revision Guide.

(b) State what you observed when the iodine was first added to each spotting tile.   **1 mark**

When the iodine was first added the mixture

....

....

> Remember to make a note of everything you observe when doing a practical investigation. This can be what you see, for example colour change or bubbles being given off, or it can be what you smell or feel.

> The rate that the amylase breaks down the starch can be calculated using:
> $$\text{rate} = \frac{1}{\text{time}}$$

(c) Calculate the mean rate that the amylase breaks down the starch for **each** pH.

Show your working. Give your answers to 3 decimal places   **3 marks**

pH 7

60 + 80 + 70 = 210

210 ÷ 3 = 70

1 ÷ 70 = ...

(d) Calculate the standard error for pH 3.   **5 marks**

Standard deviation for pH 3
Mean = 470
Sample number is 3
470 − 470 = 0 $0^2 = 0$
490 − 470 = 20 $20^2 = 400$
450 − 470 = −20 $−20^2 = 400$
0 + 400 + 400 = 800 ÷ (3 − 1) = 800 ÷ 2 = 400
$\sqrt{400}$ = 20 = standard deviation

Standard error = $\dfrac{\text{standard deviation}}{\sqrt{\text{sample number}}}$ = ...

Links For a reminder about standard deviation and standard error, look at page 78 in the Revision Guide.

(e) Plot a graph of rate of enzyme activity against pH for amylase.   **3 marks**

> Remember to write the headings and units on each axis. For 'rate' it would be 'Rate (s)'. Use as much of the graph paper as you can for the plotted area. This graph will be a bell-shaped curve. Curves should be smooth. For graphs where a line is needed, make sure it is one straight line drawn with a ruler.

Links For help with graphs of numerical data look at page 82 in the Revision Guide.

(f) Use your graph to describe any relationship between pH and amylase activity. **3 marks**

Amylase breaks down starch at pH between 3 and 10. The rate of breakdown of starch increases as the pH increases from 3 to 7 and then ....

....

....

....

....

....

....

(g) Calculate the percentage errors for the equipment you used for measuring:

    (i) the volume of each solution **1 mark**

+ or − 0.5 on measuring cylinder

0.5 × 100 ÷ 4 = 12.5%

    (ii) the temperature

> Remember to show all your working out as this can gain you marks, even if your final answer is incorrect.

The thermometer can take readings of 1°C and so has error of + or − 0.5. The temperature was controlled at a reading of 37°C ....

....

....

    (iii) the average time for pH 7 **1 mark**

The stopclock can take measurements to the nearest 0.1 second ....

....

....

(h) Explain which measurement would be most likely to affect the accuracy of the results for pH 7. **2 marks**

The percentage error was largest for the measuring cylinder. This means that each of the volumes measured can be between ....

....

....

....

Links You can see how to calculate percentage error on page 83 in the Revision Guide.

Total for Question 1 = 22 marks

Unit 3

Guided 2 A research scientist carried out a similar investigation. Their results are shown below.

pH	Mean time for starch to be broken down (s)
3	380
4	224
5	
6	92
7	96
8	
10	367
12	> 600

(a) Complete the table by suggesting average times for pH 5 and 8. **1 mark**

> Here you should suggest numbers that fit the pattern. For example for pH 5 150 seconds would be appropriate as it is a time between 222 seconds and 92 seconds.

(b) Using the secondary data explain which is the optimum pH for amylase. **2 marks**

The optimum pH is around pH 6. This is when ....

....

....

....

(c) Give **three** reasons why there is a difference between your values for time taken for starch to be broken down and those in the secondary data. **3 marks**

1. They may have used larger volumes of starch or ....

....

....

2. The amylase may not have been as fresh so ....

....

....

3. The equipment may ....

....

> Think about what the differences are. How are the data different? What might have affected this? Consider techniques and equipment used and what variables need to be controlled.

44

Links For a reminder about evaluating scientific methods, look at pages 88 and 89 in the Revision Guide.

Your colleague thinks that amylase will not act as an enzyme when the pH is 12 or higher.

(d) Comment on whether you think she is correct.

Use the secondary evidence to support your answer. **3 marks**

You must show how you have used the secondary data so make sure that you quote it within your answer.

The table shows that amylase takes ....

....

However, the secondary data mean that ....

....

....

....

Total for Question 2 = 9 marks

Unit 3
Guided

Guided

3 (a) (i) Explain why the amylase and starch solutions were kept in a water bath at 37 °C.   **2 marks**

Enzymes are affected by temperature so changes would ....

This question is about control of variables.

....

37 °C is ....

....

(ii) Explain how one other variable was controlled in this experiment.   **2 marks**

This question only asks for one other control variable. It is not enough to say keep it the same, you have to say how you keep it the same. So for volume this is done by measuring the same volume each time with a measuring cylinder.

The volume of amylase had to be controlled which in turn controlled the concentration of

the ....

This was done by..

....

....

....

(b) Explain two ways you could extend this investigation to improve the reliability of your conclusions.   **4 marks**

1. Repeat at different temperature for all pH to test if ....

This is about extending the investigation and not just carrying out more repeats.

....

....

2. Carry out test across full range of pH, 1–14, to check if ....

....

....

Total for Question 3 = 8 marks

END OF SECTION   TOTAL FOR SECTION 1 = 39 MARKS

Unit 3
Guided

Section 2

4 Effect of surface area on diffusion

Diffusion is the process where molecules move from an area of high concentration to an area of low concentration. This allows movement of substances across organic membranes, e.g. oxygen and carbon dioxide diffusion during gas exchange in the lungs.

One example of diffusion is the movement of dye in agar jelly.

You have been asked to write a plan for an investigation into the effect of surface area of agar jelly on the rate of diffusion of a dye.

12 marks

Your plan should include the following details:
- a hypothesis
- selection and justification of equipment, techniques or standard procedures
- health and safety associated with the investigation
- methods for data collection and analysis to test the hypothesis including:
  - the quantities to be measured
  - the number and range of measurements to be taken
  - how equipment may be used
  - control variables
  - brief method for data collection analysis.

> ✏️ Your answer (your plan) must be structured in a logical way so that someone else can follow it, so it is worth writing it out in rough first to ensure that you do not miss anything out. In the space below create an outline plan for your answer. Think about how you will use the dye to show diffusion rate. One way is to time how long it takes the dye to diffuse from the outside to the centre of the cube.
>
> Make sure your answer includes comments about every one of the bullet points in the question. Tick them off as they are included.

....
....
....
....
....
....

> ✏️ Now write your full answer, using your rough plan as a guide. A hypothesis should state what you think will happen. In this investigation you are looking at the effect of surface area on rate of diffusion so your hypothesis should connect these two things. A lot of hypotheses can be written 'if x increases then y will increase/decrease'. You must also use science to justify why you think your hypothesis is correct. Complete the hypothesis below.

My hypothesis is that the bigger the agar jelly cube the longer it takes for the dye to diffuse into the middle. This is because ....
....
....
....

Unit 3
Guided

....

....

> ✏️ Now consider your equipment. For this investigation you are going to need cubes of agar jelly of different sizes. How many sizes do you think you will need? How many of each size will you plan to use? List all of the equipment, including items such as beakers and measuring cylinders below.

Equipment

Ruler to measure the size of the cubes.

....

....

> Make sure you are specific about your equipment – what sizes of beakers, measuring cylinders and agar cubes will you use? How will you measure the cubes? How will you time diffusion rate?

....

....

....

....

> ✏️ Now consider the method you will use. A good method looks like a set of instructions in a recipe book.
>
> Think about your control variables. What do you need to keep the same every time? For example, concentration and volume of dye used. Make sure you also include the dependent and independent variables. Do not forget to include any health and safety procedures.

Method

Use a measuring cyclinder to add .... to a beaker

Select the first cube and…

....

....

....

....

....

> To produce a good response, you need to do more than just write down a plan, you need to explain and justify the method, equipment and techniques used. So state why you would carry out each step, why you have chosen the range and number, why you have used the equipment and so on.

....

....

....

....

> Once you have finished writing your answer, re-read it and check that it follows a logical path and could be followed by another person. Make sure you have explained and justified everything you have said.

> 🔗 **Links** Page 86 in the Revision Guide will help you to write a plan for an investigation.

Total for Question 4 = 12 marks

5 Sally investigates the effect of temperature on the rate of diffusion across a membrane.

Sally uses the following method:
- Fill dialysis tubing with starch solution at room temperature.
- Place the dialysis tubing in a beaker of iodine at room temperature.
- Start a stopclock and time when the starch solution has turned blue.

Sally repeats the steps using starch and iodine solutions at different temperatures.

The diagram shows the setup of the equipment

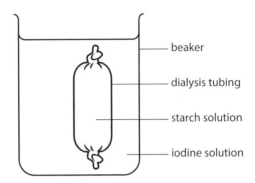

The results of Sally's investigation are shown in the table.

Temperature (°C)	Time for starch solution to turn blue (s)
20	42.12
30	37.6
43	32.24
50	25.5
66	22.78
70	22.8

Sally concludes that the higher the temperature the faster the rate of diffusion. She also concludes that over 70 °C the rate will not increase any more as the diffusion has reached its maximum rate.

Evaluate Sally's investigation.

Your answer should include reference to:
- the method of the experiment
- the results collected
- the conclusion made.

8 marks

Unit 3
Guided

> Remember to cover all of the bullet points in your evaluation of Sally's investigation. First of all, a good method is one that someone else can follow. Think of the questions you would have to ask Sally in order to follow it properly. Think of any gaps in Sally's investigation that you should mention in your evaluation, and note them down as part of the plan.

Method

....
....
....
....

> What are the variables and how are these controlled? It is also not clear that the temperature is maintained throughout the experiment using a water bath so the temperature of the solutions may cool down once they are in the beaker. Can you think of other questions about the method?

....
....
....
....
....
....
....
....

> Next you need to look carefully at Sally's results and think about these and how the results may be affected by any problems you have spotted with the method. Make some brief notes in the space below to help you plan your response.

Results

....
....

> The concentration of each solution is not given. Why do these control variables need to be kept the same throughout? If they are not, then you cannot compare the results. There are no repeats of the readings and the results are recorded to different degrees of precision. You would need to comment in your evaluation on why this is not good practice. Think about anomalies and making comparisons.

....
....
....

....
....
....
....
....
....
....

> ✏️ Finally you need to consider Sally's conclusions, whether these are supported by the experimental results and how reliable the conclusions are given any comments you have made about the method and results. Make sure you explain all the points you make.

<u>Conclusions</u>

Conclusion that higher the temperature the faster the rate of diffusion is supported by the data.

However, ...

....
....
....
....
....
....
....
....
....
....
....
....
....

Total for Question 5 = 8 marks

END OF SECTION — **TOTAL FOR SECTION 2 = 20 MARKS** — **TOTAL FOR TASK 1 = 59 MARKS**

Revision task 2

Revision task brief

You are a trainee electrician and need to understand how the brightness of a bulb in a circuit is dependent upon the power output from the bulb. Power can be calculated by measuring the voltage and current across the bulb in a circuit.

> **Links** For a reminder about calculating power look at page 72 in the Revision Guide.

Practical investigation method

You must observe safe practice when carrying out the practical investigation. This method investigates the effect of voltage on the power output of the bulb.

1. Set up the circuit as in the diagram.

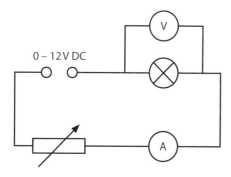

2. Set the variable power supply to 2 V. Switch on the power supply. Use the variable resistor to adjust the potential difference to 2 V.

3. Record the ammeter readings.

4. Repeat for 4 V, 6 V, 8 V, 10 V and 12 V.

5. Repeat each voltage three times.

6. Calculate the power of the bulb at each voltage.

> So that you do not have to carry out any practical work in order to revise using this Workbook, we have provided some sample results for you. You should use these to answer the rest of the questions.

Investigation results

2 V – 1.8, 1.9, 2.1 amps
4 V – 3.7, 3.7, 3.8 amps
6 V – 5.9, 6.0, 9.3 amps
8 V – 7.6, 7.7, 7.8 amps
10 V – 8.4, 8.6, 8.8 amps
12 V – 9.4, 9.2, 9.3 amps.

> **Links** For a reminder about symbols used in electrical circuits, look at page 71 in the Revision Guide.

Unit 3

> You must complete ALL activities in Section 1 and Section 2.

Section 1

1 (a) Record the experimental results in a suitable table, using the space provided. **3 marks**

 For a reminder about drawing scientific tables, look at page 76 in the Revision Guide.

> You will need to remember units and labels. You should add a mean column. Also remember that it is important to exclude results which may appear anomalous from the mean calculations.

(b) State one safety consideration you observed during the investigation. **1 mark**

 For a reminder about risk assessment during practical work, look at page 52 in the Revision Guide.

....
....

(c) Calculate the power for the bulb at each voltage.

Show your working.

2 marks

(d) Calculate the average resistance across the bulb at each voltage.

voltage = current × resistance.
V   =   A   Ω

Show your working.

Give your answer in standard form.

5 marks

> Make it clear which answer matches which voltage. Ensure you show how to rearrange the equation.
> Standard form is where just the digits are written with the decimal point after the first digit. This is followed by × 10 to the power that would put the decimal point back where it should be. So the answer for 10 V is 0.86 Ω. This in standard form will be 8.6×10^{-1}.

(e) Plot a graph of power (W) against voltage (V) `3 marks`

(f) Use your graph to describe the relationship between the brightness of the bulb and the voltage across the bulb. `3 marks`

> **Links** For a reminder about interpreting graphs, look at page 83 in the Revision Guide.

....
....
....

Remember to link brightness to power. Use data from the graph to describe the relationship

....
....
....

(g) Calculate the percentage errors for the equipment you used for measuring:

   (i) the average current at 10 volts assuming that the ammeter measures to the nearest 0.1 amp `1 mark`

....
....

(ii) the voltage at 10 volts assuming that the voltmeter measures to the nearest volt. **1 mark**

....

....

(h) Explain which measurement would be most likely to affect the accuracy of the results at 10 volts. **2 marks**

....

....

....

....

(i) Calculate the work done at 10 volts if the circuit is on for 30 seconds.

$$\text{power (W)} = \frac{\text{work done (J)}}{\text{time (s)}}$$

Show your working. **2 marks**

.... J

Total for Question 1 = 23 marks

Unit 3

2 A research scientist carried out a similar investigation. Their results are shown below.

Voltage (V)	Power (W)
2	4.21
4	15.30
6	38.76
8	
10	87.01
12	

(a) Complete the table for 8 V and 12 V. [1 mark]

Think about the results in Section 1 – the results here should follow a similar pattern.

The cost of electricity is calculated by working out how much energy has been used in kilowatt hours, kWh.

Kilowatt hours can be calculated using the following equation:

Energy transferred (kWh) = power (kW) × time (h)

Cost of 1 kWh is 10 pence.

(b) Calculate the cost of using the research scientist's bulb for 600 minutes when the voltage was 4 V. Give your answer to two significant figures.

Show your working. [3 marks]

 Links For a reminder about significant figures, look at page 77 in the Revision Guide.

You will have to convert the watts into kilowatts and the time from minutes into hours before you can work out the kWh used.

.... pence

57

(c) Give **two** reasons why there is a difference between your values for power at each voltage and those in the secondary data.

2 marks

....

....

....

> Think about the things that affect resistance and the bulb used.

....

....

> Your colleague thinks that voltage across the light bulb is directly proportional to current.

(d) Comment on whether you think she is correct.

Use your primary evidence and the secondary evidence to support your answer.

3 marks

> You should use the number of marks as a guide for how many points to make. The command word 'comment' means you need to make a judgement.

....

....

....

....

....

....

Total for Question 2 = 9 marks

3 (a) (i) Explain why the wires used in the circuit were always the same length, diameter and material. *2 marks*

> Think about the effect that the wire has on the current.

....

(ii) Identify one other variable in this investigation and how it could be controlled. *2 marks*

....

(b) Explain two ways you could extend this investigation to improve the reliability of your conclusions. *4 marks*

> This is asking how you can gain the same conclusion in other ways. This wouldn't include just repeating the experiment.

....

Total for Question 3 = 8 marks

END OF SECTION — TOTAL FOR SECTION 1 = 40 MARKS

Section 2

4 | **Effect of nitrogen, potassium and phosphorus on plant growth**
Some fertilisers contain nitrogen, potassium and phosphorus. The makers of the fertilisers claim that these three minerals improve plant growth.

You have been asked to write a plan for an investigation into the effect of mineral-containing fertilisers on the growth of plants.

12 marks

Your plan should include the following details:
- a hypothesis
- selection and justification of equipment, techniques or standard procedures
- health and safety associated with the investigation
- methods for data collection and analysis to test the hypothesis including:
  - the quantities to be measured
  - the number and range of measurements to be taken
  - how equipment may be used
  - control variables
  - brief method for data collection analysis.

Make sure you consider each prompt in the question and answer them fully. You can use extra paper if you need to, e.g. to write your rough plan in advance of your final plan. The plan must be logical and the hypothesis explained with science.

5   Jiang investigates the effect of water on plant growth.
Jiang follows the following method:
- Fill five pots with soil.
- Place a mung bean in each pot.
- Add different amounts of water to each pot every day.
- Measure the heights of the plants on the last day.

The diagram shows the setup of the equipment.

The results of Jiang's investigation are shown in the table.

Volume of water added each day (cm³)	Height of plant (cm)
20	30.5
30	32
45	38.4
60	41.65
70	42.4
80	43.8
100	39.3

Jiang concludes that the higher the volume of water added the more the plant grows. She also concludes that 80 cm³ of water will give the best plant growth.

Evaluate Jiang's investigation.

Your answer should include reference to:

- the method of the experiment
- the results collected
- the conclusion made.

8 marks

Again remember to answer all the prompts and explain each statement you make. 'Evaluate' means you must come to a judgement and you must support that judgement using the information and data. Suggestions should be given to improve the experiment to improve the conclusion.

....

....

END OF PAPER

TOTAL FOR SECTION 2 = 20 MARKS
TOTAL FOR PAPER = 60 MARKS

Unit 5: Principles and Applications of Applied Science II

Your exam

Unit 5 will be assessed through an exam, which will be set by Pearson. You will need to use your deeper understanding of core science concepts to respond to questions that require short and long answers.

Your Revision Workbook

> This Workbook is designed to **revise skills** that might be needed in your exam. The details of the actual exam may change, so always make sure you are up to date. Ask your tutor or check the **Pearson website** for the most up-to-date **Sample Assessment Material** to get an indication of the structure of your exam and what this requires of you.

To support your revision, this Workbook contains revision questions to help you revise the skills that might be needed in your exam. The revision questions are divided into three sections:

- Section A: Biology (Organs and systems)
- Section B: Chemistry (Properties and uses of substances)
- Section C: Physics (Thermal physics, materials and fluids).

> The periodic table of elements is on page 162 and the formulae sheet can be found on page 164.

Each of the sections will contain a range of different question types, including multiple choice, short answer, calculations, drawing questions and open response. You should make sure you understand what each different command word is asking you to do in the question.

> To help you revise skills that might be needed in your Unit 5 exam this Workbook contains two sets of revision questions starting on pages 65 and 86. See the introduction on page iii for more information on features included to help you revise.

> - Read each question carefully before you start to answer it.
> - Check your answers if you have time at the end.

Revision test 1

Section A: Organs and systems

Answer ALL questions. Write your answers in the spaces provided.

1 A young man has heart palpitations, feels dizzy and is short of breath. At hospital he has an electrocardiogram (ECG), which indicates he has tachycardia.

(a) Choose the ECG trace below which is most likely to belong to the young man. **1 mark**

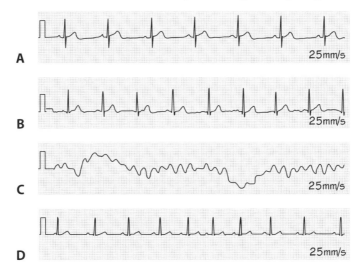

☐ A
☐ B
☐ C
☐ D

> You need to remember which is a slow, fast or irregular rhythm.

Links You can revise the cardiac cycle and ECGs on page 92 of the Revision Guide.

> Guided

(b) Describe what tachycardia means. **1 mark**

Tachycardia is when the heart is beating faster than normal, at more than

....

(c) The young man's normal resting heart rate was about 62 bpm, but on this occasion it was 108 bpm.

Calculate the % increase in heart rate.

3 marks

> When doing calculations, it is good to roughly estimate your answer. The young man's heart rate has not quite doubled, but it has increased by more than a half, so the answer should be more than 50% and less than 100% (100% is double).

His heart rate increased by 108 − 62 = ... bpm

% increase = $\dfrac{....................}{62}$ × 100 = ... %

> Note you are calculating the increase from the original or normal, so divide by 62 not 108.

(d) One of the problems associated with tachycardia is that the ventricles do not have time to fill properly.

Name the stage in the cardiac cycle when the ventricles are contracting.

1 mark

....

> Remember diastole is a recovery phase, systole is a contraction.

Links You can revise the heart and the cardiac cycle on pages 90 and 92 of the Revision Guide.

(e) Explain two of the young man's symptoms.

4 marks

The young man feels 'dizzy' because not enough oxygen-rich blood is being pumped to the

....

The young man is 'short of breath' because not enough blood is being pumped to the lungs

to exchange carbon dioxide with oxygen. The body's response to this is to

....

Total for Question 1 = 10 marks

2 A woman had pneumonia as a child. This resulted in some damage to the alveoli.

(a) In which organ would you find alveoli tissue? **1 mark**

☐ **A** lungs

☐ **B** small intestine

☐ **C** liver

☐ **D** brain

(b) State the name of the structures labelled A–C. **3 marks**

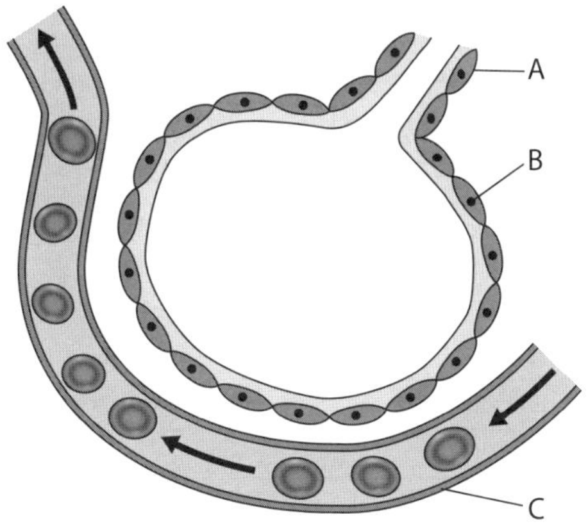

A ....

B ....

C ....

Links You can revise alveoli on page 95 of the Revision Guide.

(c) Explain two ways the structure of an alveolus is linked to improved diffusion. **4 marks**

Alveoli have thin walls ....

....

....

....

....

Make sure that you have stated the function of alveoli, and then say how each of the two features help the alveoli achieve the function.

(d) The damage caused by the pneumonia was 'scarring' of the alveoli, which reduced the number of 'working' alveoli.

Explain how this damage to the alveoli could affect the woman's running. **3 marks**

Fewer 'working alveoli' means a reduced surface area of the lungs. This means that less ....

....

....

....

The woman is a keen marathon runner and always performed well, despite the damage to her lungs. Some tests were carried out to investigate this as part of a student project.

	Forced vital capacity (dm^3)	Breathing rate at rest (breaths per minute)	Stroke volume (ml)	Heart rate (bpm)	Cardiac output (ml/minute)
Normal	4.0	12	70	62	
Female marathon runner	3.2	18		51	4590

Links You can revise measuring lung volumes on page 97 of the Revision Guide.

(e) Complete the table above.

Show your working. **4 marks**

cardiac output = heart rate × stroke volume

Links You can revise cardiac output on page 90 of the Revision Guide.

(f) The diagram shows a spirometer reading.

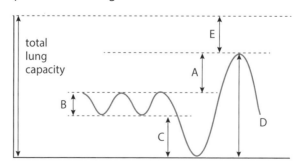

State the letter that represents the forced vital capacity. **1 mark**

....

Forced vital capacity is when you inhale and exhale as deeply as possible.

Unit 5
Guided

 (g) Explain the biological adaptations that the woman's body has made to compensate for her reduced lung capacity. The woman has an increased stroke volume and cardiac output. Use the other information in the table to help you. **3 marks**

> In this question, you are asked to use the information in the table, so you should go through everything that is measured. Compare the female marathon runner to the other person and suggest how the difference might help make up for her reduced lung capacity.

The female marathon runner's body seems to have adapted to provide more oxygen into her body by increasing her rate of breathing from 12 ....

....

....

....

....

....

....

> For the comparison to be valid the other person would need to be the same age, sex, weight, a non-smoker and otherwise healthy. The question does not ask this, but it could!

 You can revise lung ventilation and measuring lung capacity on pages 96 and 97 of the Revision Guide.

Total for Question 2 = 19 marks

Unit 5
Guided

3 A GP received the results back from some urine tests that had been carried out for some of his patients. The table below shows the results.

Patient	Protein (g/dm³)	Glucose (g/dm³)
A	0.100	0.00
B	0.003	0.05
C	0.006	0.20

(a) Patient B appears to be healthy, although there is a trace of glucose present. Suggest a reason why this result is not 0.00 g/dm³ as recorded for Patient A. **1 mark**

> Remember that having glucose in the urine can indicate diabetes.

....

(b) Patient C has an abnormal result. Describe what is abnormal about it. **1 mark**

> To answer this question, look at the results for patient A. Which substance is similar to another patient and which is different?

....

....

....

Guided

(c) Patient A has an abnormally high protein reading. Explain why protein is not normally found in urine and why it might be a concern. **3 marks**

> Think about what you know about protein molecules.

Proteins usually remain in the blood because the molecules are too ....

....

....

....

....

....

....

> Glucose is an essential molecule and should be reabsorbed into the blood.

(d) Explain why some molecules are reabsorbed and others are not, and how they are reabsorbed. Your answer should refer to the loop of Henle.

6 marks

> This is an extended writing question, so all your points need to be in a logical order.
>
> Use these questions to help you:
> - Which substances are reabsorbed? Why?
> - Which substance is excreted? Why?
>
> Now think about the loop of Henle where substances are reabsorbed, draw a diagram if this helps you. An annotated diagram will also score marks.
> - Which substance is absorbed first? Why and how?
> - How does this affect the filtrate concentration?
> - What substances are absorbed next? Why and how?

Total for Question 3 = 11 marks

END OF SECTION — TOTAL FOR SECTION A = 39 MARKS

Section B: Properties and uses of substances

Answer ALL questions. Write your answers in the spaces provided.

1. Propane, C_3H_8, and propene, C_3H_6, are both hydrocarbons.
   They each belong to different homologous series.

   (a) (i) Give the general formula of the homologous series to which **propane** belongs. **1 mark**

   C_nH ....

   (ii) Give the **name** of the homologous series to which **propene** belongs. **1 mark**

   ....

   *Think about the ending of the name of the molecule – pro**pene**.*

   (b) Propane is a saturated hydrocarbon, while propene is unsaturated.

   (i) State what is meant by the term *unsaturated*. **1 mark**

   ....
   ....

   You need to consider the type of bonds between the carbon atoms in the hydrocarbons.

   (ii) Describe what would be observed on adding bromine water to both a saturated and an unsaturated hydrocarbon. **2 marks**

   With a saturated hydrocarbon the bromine water stays ....

   With an unsaturated hydrocarbon the bromine water turns ....

   Propane reacts with chlorine in the presence of UV radiation to form a mixture of chlorinated products.
   This is one possible reaction:

   $$C_3H_8 + Cl_2 \rightarrow C_3H_7Cl + HCl$$

   (c) Write equations for each of the steps of the mechanism for this reaction.

   Step 1 is given for you. **3 marks**

   Step 1 (initiation) $Cl_2 \rightarrow 2Cl\cdot$

   Step 2 (propagation – two equations are required)

   Equation 1 ....

   Equation 2 ....

   Step 3 (termination) ....

   Initiation leads to the formation of chlorine radicals. Propagation involves a chain reaction. Termination removes radicals from the reaction mixture.

Propene undergoes an electrophilic addition reaction with bromine.

⟶   ⟶ H—C—C—C—H
                     | | |
                     Br Br H
                     H H H
                     | | |

(with H, H, H on top of the right-hand structure)

(d) Some parts of the mechanism above are missing.

Draw the intermediates to complete the equation. Use curly arrows in your answer.   **4 marks**

A curly arrow shows the movement of a pair of electrons. This could be a pair of electrons in a covalent bond or a non-bonding (lone) pair of electrons.

Links You can revise reactions of alkanes and alkenes on pages 113 and 114 of the Revision Guide.

Total for Question 1 = 12 marks

Unit 5
Guided

2 Bauxite contains aluminium hydroxide, Al(OH)$_3$, and several impurities such as iron(III) oxide, Fe$_2$O$_3$, and titanium(IV) oxide, TiO$_2$.

To remove the impurities, powdered bauxite is mixed with sodium hydroxide solution and the mixture is then heated. The aluminium hydroxide reacts but the impurities do not.

Guided

(a) (i) Write a chemical equation for the reaction between sodium hydroxide and aluminium hydroxide.

State symbols are not required. **2 marks**

.... NaOH + ... → Na$_3$Al(OH)$_6$

(ii) Explain why **only** the aluminium hydroxide reacts with the sodium hydroxide. **3 marks**

You need to give a reason why aluminium hydroxide reacts with sodium hydroxide, **and** a reason why both iron(III) oxide and titanium(IV) oxide do not.

Sodium hydroxide is a ....

....

....

....

....

....

....

The iron(III) oxide and titanium(IV) oxide are removed by filtration and aluminium hydroxide is precipitated from the filtrate.

(b) (i) State how aluminium hydroxide is precipitated from the filtrate. **1 mark**

....

....

The aluminium hydroxide precipitate is then heated to convert it into alumina.

(ii) Write a chemical equation for the action of heat on aluminium hydroxide.

State symbols are not required. **1 mark**

....

74

Titanium is another metal that is extracted from its ore, which contains titanium(IV) oxide, TiO_2. The two main stages in the extraction are represented by these equations.

Stage 1 $TiO_2 + 2Cl_2 + C \rightarrow TiCl_4 + CO_2$ (1275 K)

Stage 2 $TiCl_4 + 2Mg \rightarrow Ti + 2MgCl_2$ (1300 K in an atmosphere of argon)

(c) (i) Explain why Stage 2 is carried out in an atmosphere of argon and not air. **2 marks**

You need to consider what would happen if either titanium or magnesium were heated in air, and what would happen if either metal were heated in argon.

(ii) Explain why the reaction in Stage 2 is described as a redox reaction. **4 marks**

Work out the oxidation numbers of each element involved in the reaction. Identify the two elements whose oxidation numbers have changed. Remember that an increase in oxidation number is oxidation and that a decrease in oxidation number is reduction.

Links You can revise aluminium and titanium on page 107 of the Revision Guide.

Total for Question 2 = 13 marks

Unit 5
Guided

3 The formation of substances is accompanied by enthalpy changes.

(a) Write an equation, including state symbols, representing the standard enthalpy change of formation of **solid** lead(II) oxide (PbO). **(2 marks)**

> You must include state signs in the equation, and the elements must be in their standard states under standard conditions.

....

(b) State what is meant by the term *standard enthalpy change of formation*. **(3 marks)**

> Include the conditions of temperature and pressure in your answer. Remember that the standard enthalpy change of formation is for one mole (1 mol) of substance formed.

....

....

....

....

....

Some metal priming paints contain 'red lead' (Pb_3O_4). This can be made by heating lead(II) oxide in the presence of air.

$$3PbO(s) + \tfrac{1}{2}O_2(g) \rightarrow Pb_3O_4(s)$$

The table shows the standard enthalpy changes of formation, $\Delta_f H^\ominus$, of PbO(s) and Pb_3O_4(s).

	$\Delta_f H^\ominus$ / kJ mol^{-1}
PbO(s)	−219
Pb_3O_4(s)	−735

Guided

(c) Use the values in the table to calculate the standard enthalpy change ($\Delta_r H^\ominus$) for the above reaction. **(4 marks)**

> First obtain an expression for $\Sigma \Delta_f H^\ominus$(reactants); this will include the unknown $\Delta_r H^\ominus[PbO_2(s)]$. Then calculate $\Sigma \Delta_f H^\ominus$(products). Finally use $\Delta_r H^\ominus = \Sigma \Delta_f H^\ominus$(products) − $\Sigma \Delta_f H^\ominus$(reactants) and rearrange to make $\Delta_r H^\ominus[PbO_2(s)]$ the subject of the equation.

$\Sigma \Delta_f H^\ominus$(reactants) = ...

$\Sigma \Delta_f H^\ominus$(products) = ...

$\Delta_r H^\ominus = \Sigma \Delta_f H^\ominus$(products) − $\Sigma \Delta_f H^\ominus$(reactants) = ...

$\Delta_r H^\ominus[PbO_2(s)]$ = ...

= .... kJ mol^{-1}

Links You can revise enthalpy changes on pages 116–120 of the Revision Guide.

Total for Question 3 = 9 marks

4 Sodium hydroxide is manufactured by the electrolysis of brine.

Three different types of cell are used:
- the flowing mercury cathode cell
- the diaphragm cell
- the membrane cell.

	Mercury	**Diaphragm**	**Membrane**
Construction costs	expensive	relatively cheap	relatively cheap
Operation	toxic mercury must be removed from the waste material before it is discarded	diaphragm has to be replaced frequently	low maintenance costs
Product	high purity and produces the required concentration of sodium hydroxide (i.e. 50%)	low purity and produces a very low concentration of sodium hydroxide that requires evaporation	high purity but concentration of sodium hydroxide is lower than required (i.e. 35%)
Overall energy consumption/ megawatts per tonne of chlorine	3.3–3.5	3.3–3.5	2.8–3.0

Discuss the most suitable method for manufacturing sodium hydroxide.

6 marks

Include a reference to each of the following:
- The relative cost of making each cell.
- Any environmental impact that each process may have.
- Any maintenance costs involved.
- The purity of the sodium hydroxide produced.
- The relative cost of electricity for each process.

Finally, reach a conclusion as to which method is the best.

Unit 5
Guided

....
....
....
....
....
....
....
....

Links You can revise the electrolysis of brine on page 108 of the Revision Guide.

Total for Question 4 = 6 marks

END OF SECTION — TOTAL FOR SECTION B = 40 MARKS

Section C: Thermal physics, materials and fluids

Answer ALL questions. Write your answers in the spaces provided.

1 The picture shows a bomb calorimeter.
It can be used to determine the calorific value of a sample of fuel.
The reaction is contained at a constant volume regardless of pressure changes.

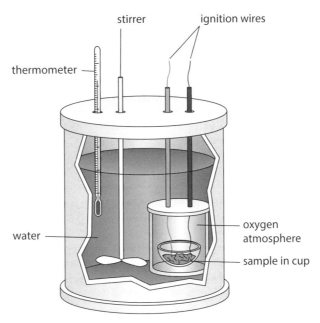

After igniting the fuel, the temperature of the water in the surrounding bath rose from 20.2 °C to 23.1 °C

(a) (i) Explain whether work has been done by the reaction. **2 marks**

No work has been done because $W = p\Delta V$ and there has been no change in ....

....

(ii) Explain the reason for the **sign** of the change in internal energy, ΔU. **3 marks**

> You need to apply the First Law of Thermodynamics: $\Delta U = Q - W$. Think about the size and sign of each quantity and consider the direction of the heat energy transfer across the system boundary.

ΔU is ...., as $W = 0$ and Q is ....,

because ....

> The water bath contained 2.40 litres of water.
> The density of water at that temperature is 1000 kg m^{-3}
> The specific heat capacity of water is 4.18 kJ kg^{-1} K^{-1}
> The heat capacity of the steel bomb itself and the water bath vessel is 1.25 kJ K^{-1}.

Guided (b) Calculate ΔU for the fuel/oxygen system in the burning process.   **3 marks**

The heat capacity of the water is:

2.40×10^{-3} m^3 × 1000 kg m^{-3} × 4.18 kJ kg^{-1} K^{-1} =
.... kJ K^{-1}

> Do not forget to write in the sign of the answer and its units.

So the total heat capacity of the calorimeter, water bath and water

= .... + 1.25 kJ K^{-1} = ... kJ K^{-1}

$\Delta U = Q = -$ .... kJ K^{-1} × (23.1 °C − 20.2 °C)

= ....

....

(c) Explain one impact on the result obtained if the fuel had been burned at atmospheric pressure.   **3 marks**

> What would have happened to the volume of the gas when it was heated by the combustion process? What effect would that have on the quantities in the First Law of Thermodynamics equation, $\Delta U = Q - W$?

....

....

....

....

....

....

 Links You can revise The First Law of Thermodynamics on page 124 of the Revision Guide.

Total for Question 1 = 11 marks

2

Dry ice (solid carbon dioxide) is sometimes used to keep things cool.

The diagram shows the stages of producing solid carbon dioxide pellets.

Liquefied CO_2 under high pressure enters an atmospheric pressure chamber, where sudden expansion causes both boiling and freezing to occur at the same time.

Around 2.3 to 2.5 kg of liquid CO_2 produces 1 kg of dry ice at a temperature of −78 °C.

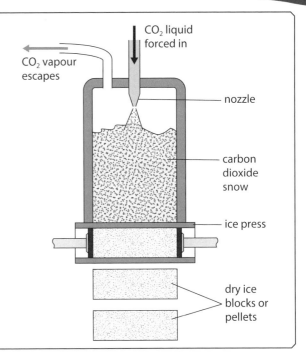

(a) Give the temperature of the system in K.

.... K

> Remember that absolute zero (0K) is −273.15 °C.

> When a liquid boils, the process of change of state to a gas requires the latent heat of vaporisation to be supplied.

Guided (b) Explain why expansion and boiling of liquid CO_2 causes rapid cooling. **2 marks**

> The First Law of Thermodynamics can be written as $Q = \Delta U + W$, where Q is the heat input to the system and W is the work done by the system.

The supply of energy for the latent heat of vaporisation comes from the sensible heat rejected as

....

....

Guided (c) Applying the First Law of Thermodynamics to the process of compressing and liquefying CO_2 gas, explain what is the sign of ΔU and of W. Hence comment on the sign and size of Q. **5 marks**

Q is zero (no external heat source) and expansion means work is done ($p\Delta V$), so ΔU is

.... and the temperature falls.

Boiling removes .... of vaporisation which makes the temperature fall further.

> **Links** You can revise absolute zero on page 123 of the Revision Guide.

Total for Question 2 = 7 marks

3 The table compares the mechanical properties of various structural materials.

Structural materials	Young's modulus (GPa)	Ultimate tensile strength (MPa)	Yield strength (MPa)	Elongation at break	Density (kg m⁻³)
ABS plastics	2.3	42	40	25%	1050
Aluminium	69	110	95	15%	2700
Steel, stainless	180	860	502	70%	7870
Concrete, high-strength	30	40 (compression)	–	–	2000
Borosilicate glass	68–81	2000 (compression)	–	–	2200
Polyethylene HDPE (high density)	0.8	37	26–33	150%	930–970

(a) State which is the most ductile material. **1 mark**

> Choose between the materials that show a yield point and an elongation before failure.

....

(b) Explain which of the materials listed is the most elastic. **2 marks**

....

....

....

> Which property is an elastic modulus, telling you about the strength of the spring back?

....

(c) Explain why concrete and glass are tested in **compression** as opposed to tension. **2 marks**

....

....

....

> What kind of failure would you expect for these materials?

....

Aluminium and stainless steel are widely used in making car body panels.

(d) (i) Explain which **one** of the strength values should be used by engineers for design calculations on these parts. 2 marks

Engineers should use .... because body panels should

not be permanently .... by normal stresses.

(ii) Explain which of the materials has the best strength to weight ratio. 4 marks

Calculating (yield strength / density) for:

ABS plastics = 40 / 1050 MPa m^3 kg^{-1} = ...

Aluminium = ... / .... MPa m^3 kg^{-1} =

....

Stainless steel = ... / .... MPa m^3 kg^{-1} =

....

So, ...

....

Links — You can revise shape change on page 130 of the Revision Guide.

Total for Question 3 = 11 marks

4 The sketch graph shows the viscous behaviour of some liquids and semi-solid materials when shear stresses are applied to them.

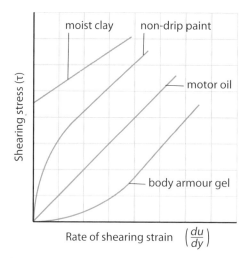

(a) State what is meant by shear stress. **1 mark**

....

(b) Explain **two** features of a Newtonian fluid. **4 marks**

....

....

....

....

....

....

....

(c) Compare the properties of the materials shown in the graph. **6 marks**

The effective viscosity of a fluid is ....

....

So a steep straight line would mean ....

....

and a shallow straight line would mean ....

....

....

....

....

Non-drip paints have ....

....

....

....

....

....

....

Motor oils need to maintain a steady viscosity so that ....

....

Links You can revise non-Newtonian fluids on page 133 of the Revision Guide.

Total for Question 4 = 11 marks

END OF SECTION
END OF PAPER

TOTAL FOR SECTION C = 40 MARKS

TOTAL FOR PAPER = 120 MARKS

Revision test 2

Section A: Organs and systems

Answer ALL questions. Write your answers in the spaces provided.

1 When a person smokes a cigarette, one of the gases that is inhaled is carbon monoxide.

(a) Explain why diffusion is passive. **1 mark**

....

(b) Explain why carbon monoxide will diffuse from the alveoli into the blood. **2 marks**

.... Diffusion always happens
.... down a concentration gradient.

....

....

....

🔗 **Links** You can revise the human lungs on page 95 of the Revision Guide.

(c) Give three reasons why gases can diffuse through the alveoli membrane into the capillaries. **3 marks**

....

....

....

....

....

🔗 **Links** You can revise passive transport on page 104 of the Revision Guide.

Unit 5

A special type of diffusion is osmosis. This occurs in the kidney tubule.

(d) Identify the substance that diffuses via osmosis. **1 mark**

....

Osmosis is the diffusion of only one type of molecule.

(e) Antidiuretic hormone (ADH) is released by the pituitary. Explain the role of ADH. **2 marks**

....
....

ADH is a hormone that is important in osmoregulation.

....
....
....

Links You can revise osmoregulation on page 100 of the Revision Guide.

Total for Question 1 = 9 marks

Unit 5

2 The water flea *Daphnia* is used to investigate the effects of various substances such as caffeine on heart rate. Some students wanted to measure the effect of nicotine on heart rate and used the water flea *Daphnia*.

(a) Give **two** reasons why *Daphnia* is more suitable than human experimentation. **2 marks**

....

....

....

....

> Be careful to choose two reasons why *Daphnia* **should** be used, rather than reasons why they should not be used.

....

....

(b) Describe how the students could investigate the effect of nicotine on the heart rate of *Daphnia*. **4 marks**

....

....

....

....

> Remember *Daphnia* should be in their test environment and need to be viewed and measured.

....

....

....

....

....

....

(c) Name three variables the students need to control in this experiment. **3 marks**

....

....

....

....

....

....

(d) Before they started the experiment to test the effect of heart rate on *Daphnia*, the students needed to know the **range** of concentration of nicotine solutions to use. Give a suggestion as to how they might have done this. **2 marks**

> This is a preliminary experiment to work out the parameters of the main experiment.

....
....
....
....

Links You can revise the effect of caffeine on heart rate on page 94 of the Revision Guide.

Total for Question 2 = 11 marks

Unit 5

3 A new fungicide is being developed called 'Botokil'. It affects cell membranes of the fungus *Hymenoscyphus fraxinus*, which causes the disease chalara dieback of ash trees. The new fungicide is designed to work by affecting the carrier proteins in the cell membrane of the fungus, so they cannot use ATP.

Links You can revise active transport on page 105 of the Revision Guide.

(a) What role does ATP play in cells? **1 mark**

☐ **A** It communicates information between cells.

☐ **B** It provides energy to cells.

☐ **C** It contains all the information about an organism.

☐ **D** It codes for proteins.

(b) Explain how 'Botokil' will affect the fungus. **2 marks**

....

....

....

Think about what will happen if the carrier proteins do not have ATP.

....

....

(c) The phospholipid bilayer is another important part of the cell membrane structure.

Describe what is meant by 'phospholipid bilayer'. **3 marks**

....

....

You could answer using an annotated diagram.

....

....

....

....

....

....

Links You can revise cell surface membrane on page 4 of the Revision Guide.

Total for Question 3 = 6 marks

4 Around 7 million people suffer from cardiovascular disease (CVD) in the UK. CVD happens because of damage to the arteries in organs, usually caused by a build-up of fatty deposits. When this happens it is called atherosclerosis.

(a) What is **not** a treatment for CVD? `1 mark`

☐ **A** hypertensives

☐ **B** statins

☐ **C** transplant

☐ **D** dialysis.

(b) These fatty deposits do not occur in veins. Compare the structure of veins and arteries. `4 marks`

The fatty deposits accumulate in the thick wall between the endothelial lining and the smooth muscle.

....

You could answer this question using a table.

....

....

....

....

....

....

....

....

Links You can revise blood vessels and cardiovascular disease on pages 91 and 93 of the Revision Guide.

(c) State which factors should **not** be focussed on in a health campaign to reduce the levels of CVD. `3 marks`

....

Think of those factors that people cannot change.

....

....

91

(d) State the factors that should be a focus of a health campaign to reduce the levels of CVD. For each one explain what effect the change will have. **6 marks**

> Think of the lifestyle factors that are known to affect the risk of CVD, and for each one explain how improving lifestyle will help the body and reduce the risk of CVD.

....
....
....
....
....
....
....
....
....
....
....
....

Total for Question 4 = 14 marks

END OF SECTION — **TOTAL FOR SECTION A = 40 MARKS**

Section B: Properties and uses of substances

Answer ALL questions. Write your answers in the spaces provided.

1. Crude oil is refined by fractional distillation to produce more useful hydrocarbons.
   These hydrocarbons are alkenes and alkanes.
   Members of a homologous series of compounds have the same general formula.
   The general formula for alkanes is C_nH_{2n+2}.

   (a) (i) State the molecular formula of octadecane. *1 mark*

   ....

   (ii) State **two** other features of a homologous series of hydrocarbons. *2 marks*

   Think about their structure, chemical and physical properties.

   1 ....

   2 ....

   The fractions obtained from the fractional distillation of crude oil often undergo cracking.

   (b) (i) Explain **two** reasons why the cracking of long-chain alkanes such as octadecane is important. *4 marks*

   You need to think about what types of hydrocarbon are most needed by industry.

   ....

   (ii) Under certain conditions hexadecane can be cracked into octane and an alkene.
   Write a chemical equation to represent the cracking reaction. *2 marks*

   ....

> Propene is an unsaturated hydrocarbon and contains a π bond.

(c) (i) Describe, using a diagram, how p orbitals are involved in the formation of the π bond in propene. **2 marks**

....

....

> A major commercial use of propene is the production of the polymer, poly(propene).

(ii) Draw a section of poly(propene) showing **two** repeat units. **2 marks**

Links You can revise polymerisation on page 115 of the Revision Guide.

Total for Question 1 = 13 marks

2 Propane, C_3H_8, is a gas at room temperature and pressure.
It is used in blowtorches to melt bitumen when applying felt to a roof.

(a) Write a chemical equation for the complete combustion of propane.

State symbols are not required.

2 marks

....

Remember to balance the equation.

(b) Describe what is meant by the term *standard enthalpy change of combustion*.

3 marks

....

....

....

....

....

....

(c) The apparatus shown is used to determine the enthalpy change of combustion of propane.

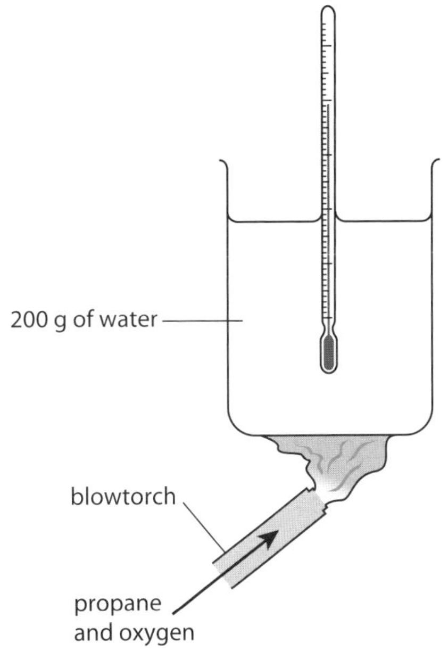

200 g of water

blowtorch

propane and oxygen

The temperature of the water changes from 19.8 °C to 70.1 °C when 1.00 g of propane is burned.
Assume the propane is burned under ideal conditions and all the energy is transferred to the water.

Change in energy = mass × specific heat capacity × change in temperature

(i) Calculate the heat energy, in kJ, transferred to the water.

Give your answer to three significant figures.

(The specific heat capacity of water is 4.18 J g^{-1} K^{-1}.)

3 marks

Heat energy = ... kJ

(ii) Give one reason why the actual energy transferred to the water is lower than that calculated.

1 mark

Values of enthalpy change of combustion can be used to calculate enthalpy changes of formation.

The equation for the enthalpy change of formation of propane is:

3C(s) + 4H$_2$(g) → C$_3$H$_8$(g)

The table gives the enthalpy changes of combustion of carbon, hydrogen and propane.

	Enthalpy change of combustion (kJ mol^{-1})
Carbon, C(s)	−394
Hydrogen, H$_2$(g)	−286
Propane, C$_3$H$_8$(g)	−2219

(d) Calculate the enthalpy change of formation ($\Delta_f H$), of propane.

Show all of your workings.

3 marks

Be careful with the signs.

$\Delta_f H$ [propane] .... kJ mol^{-1}

Links You can revise enthalpy changes on pages 116–120 of the Revision Guide.

Total for Question 2 = 12 marks

3 Electrolysis of brine is used to produce chlorine.
One of the products of the process is hydrogen.

(a) Write the equation for the reduction of water that takes place at the cathode.   **2 marks**

....

....

(b) Explain the use of one **other** chemical that is produced during the electrolysis of brine.   **2 marks**

....

....

....

....

(c) Explain one disadvantage of the process of brine electrolysis.   **2 marks**

....

....

....

....

The chlorine and hydrogen produced during the process can be recombined to make HCl.
$$Cl_2(g) + H_2(g) \rightarrow 2\,HCl(g) \qquad \Delta H = -185\,kJ\,mol^{-1}$$

(d) Explain whether this reaction is endothermic or exothermic.   **3 marks**

....

....

....

....

....

Total for Question 3 = 9 marks

4 Transition metals and their compounds act as catalysts for many important industrial chemical reactions.

Iron is a catalyst in the manufacture of ammonia (the Haber process).

Vanadium(V) oxide is a catalyst for one stage of the manufacture of sulfuric acid (the Contact process).

Compare the action of the catalysts in the Haber process and the Contact process.

You may use chemical equations, where appropriate, to support your answer.

6 marks

Total for Question 4 = 6 marks

END OF SECTION — TOTAL FOR SECTION B = 40 MARKS

Section C: Thermal physics, materials and fluids

Answer ALL questions. Write your answers in the spaces provided.

1   The graph shows how stress varies with strain for a metal alloy in the form of a wire under tension.

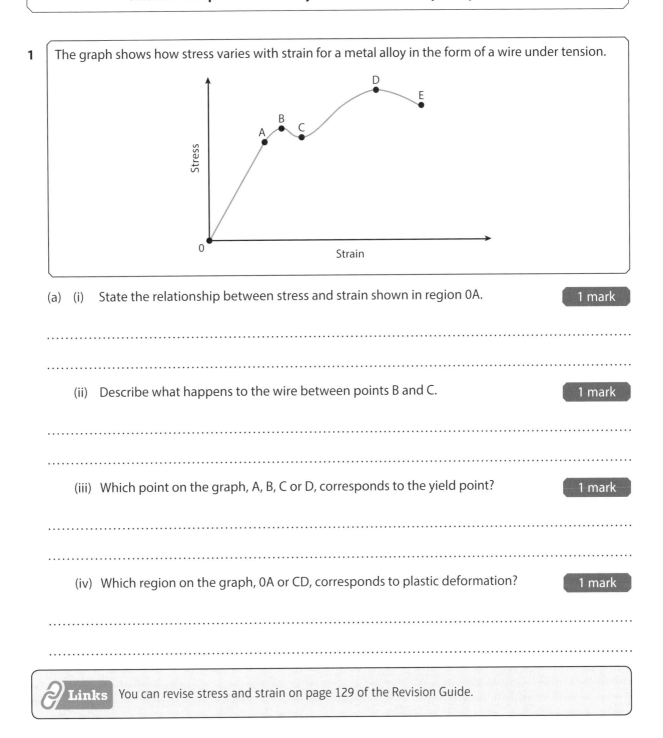

(a) (i)   State the relationship between stress and strain shown in region 0A.   1 mark

....

....

(ii)   Describe what happens to the wire between points B and C.   1 mark

....

....

(iii)   Which point on the graph, A, B, C or D, corresponds to the yield point?   1 mark

....

....

(iv)   Which region on the graph, 0A or CD, corresponds to plastic deformation?   1 mark

....

....

Links   You can revise stress and strain on page 129 of the Revision Guide.

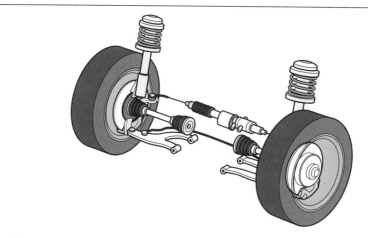

The picture shows part of a truck's suspension system. There are six springs in total. Each spring supports an equal fraction of the weight of the truck.

(b) Each spring has a spring constant of 25 000 Nm−1. The weight of the truck is 24 000 N. Calculate the amount each spring is compressed by the weight of the truck. Show all your working and state the unit in your answer.

3 marks

....

....

....

....

....

....

Compression = ...

(c) Show that the work done in compressing one of these springs is 320 J.

1 mark

Make sure you show all your working, as marks are given for the working, not the value of the answer which has been given to you.

....

....

....

....

(d) Springs sometimes fail due to creep or fatigue. State two differences in these types of failure.

2 marks

> Think about the types of loading on the springs and their environment.

....

....

....

....

....

....

Total for Question 1 = 10 marks

Unit 5

2 Hydraulic lifts are used to move heavy items safely.

car (m = 2000 kg)
platform (m = 270 kg)
piston
hydraulic fluid
P

One lift can raise a car of mass 2000 kg from an initial height of 300 mm to 1200 mm above floor height. The mass of the platform (which must also be lifted) is 270 kg.

(a) (i) State the equation to find the work done in lifting a mass. **1 mark**

....

....

(ii) Show that the work done when the car and platform are lifted from 300 mm to 1200 mm is about 20 kJ. **2 marks**

> Add the two masses to calculate the total mass being lifted. Show all your working, as marks are given for the working, not the value of the answer which has been given to you. Remember to use SI units.

....

....

....

....

....

Work done = ... J

Links You can revise work done on page 121 of the Revision Guide.

(b) (i) A hydraulic pump is used to operate the lift. The work done W by the pump can be expressed as $W = p\Delta V$, where p is the pressure exerted by the pump. Give the meaning of the symbol ΔV. **1 mark**

....

102

(ii) The hydraulic pump has a piston that moves through a volume of 800 cm³ in order to lift the car. Using the result from part (a) and the equation in (b)(i), estimate the minimum pressure produced by the hydraulic pump. **2 marks**

> Equate the 20 kJ result from part (a) to the pressure–volume work done by the piston of the hydraulic pump.

....

....

....

....

(iii) Explain why the actual pressure may be greater than that calculated in (b)(ii). **2 marks**

....

....

....

....

> The hydraulic pump is powered by an electric motor. The energy input required to lift the car and platform in (a) is 45 200 J.

(c) Calculate the overall efficiency of this lift when lifting the total mass. **2 marks**

....

> Remember that the result from (a) should be about 20 kJ.

....

....

....

....

Efficiency = ... %

Links You can revise efficiency on page 127 of the Revision Guide.

Total for Question 2 = 10 marks

3 | The illustration shows a modern car suspension system, with the damper shown below in more detail.

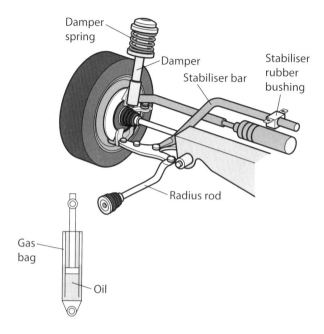

Under the weight of the car, the damper piston is pushed downwards through the oil, which flows slowly upwards through small holes in the piston. The spring returns the vehicle to its correct height above the road. The damper reduces the size and duration of oscillations.

(a) Describe the energy transfers in both the spring and damper. **2 marks**

....
....
....
....

(b) State **two** properties of a fluid that must be considered if it is to be used in damping. **2 marks**

....
....
....
....

(c) State **two** properties of a metal that must be considered if it is to be used as a damper spring. **2 marks**

....
....
....
....

(d) When a vehicle is driven, the hydraulic fluid may get hot. Describe how the First and Second Laws of Thermodynamics apply to the energy transfers affecting the fluid. **2 marks**

> For the First Law, think about whether heat has been transferred and whether work has been done. For the Second Law, consider whether the thermal energy generated can be recovered as work, so is the process reversible?

....

....

....

....

Total for Question 3 = 8 marks

Unit 5

4 A student is investigating the melting of ice.

0.8 kg of ice at 0 °C is heated in a closed container by a 1 kW heater. As the ice melts, this meltwater is warmed by the heater.

Latent heat of fusion of ice = 330 kJ kg^{-1}

Specific heat capacity of water = 4.2 kJ kg^{-1} °C^{-1}

(a) State what is meant by latent heat of fusion. **1 mark**

....

> **Links** Look at page 122 of the Revision Guide for a reminder about latent heat.

(b) Assuming no thermal energy is lost to the surroundings, calculate the temperature of the water after 5 minutes. Give your answer to the nearest degree. **3 marks**

> Start by writing down the equations you will use. List the data that are given in the question and identify the unknown value. Remember to show all your working.

....

....

....

....

....

....

....

(c) Ice is produced in a freezer. If the inside of the freezer is maintained at −18 °C and the thermal energy reservoir is at 22 °C, calculate the maximum theoretical coefficient of performance (COP) of the freezer. **2 marks**

> Remember to convert temperatures into Kelvin first.

....

....

....

....

The evaporator coils in a freezer allow the high pressure refrigerant to absorb heat from the air inside the freezer, thus cooling the freezer compartment.

(d) The table below shows data about the properties of three different materials.

Material	Strength/ MPa	Ductility (% elongation at break)	Density/g cm^{-3}	Thermal conductivity/ kW m^{-1} K^{-1}	Cost/£ per tonne
aluminium alloy	300	15	2.7	205	1260
copper	220	56	8.9	401	5000
brass	500	32	8.7	109	2500

Comment on which of the materials in the table is most suitable for making evaporator coils.

6 marks

Total for Question 4 = 12 marks

END OF PAPER

TOTAL FOR SECTION C = 40 MARKS
TOTAL FOR PAPER = 120 MARKS

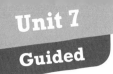

Unit 7: Contemporary Issues in Science

Your set task

Unit 7 will be assessed through a task, which will be set by Pearson. You will need to use your knowledge of contemporary science issues and their impact on the world we live in to understand a contemporary scientific issue described in articles provided to you. You will then answer questions that will require you to analyse and interpret the articles and justify your answers.

Your Revision Workbook

> This Workbook is designed to **revise skills** that might be needed in your assessed task. The details of the actual assessed task may change, so always make sure you are up to date. Ask your tutor or check the **Pearson website** for the most up-to-date **Sample Assessment Material** to get an indication of the structure of your assessed task and what this requires of you.

To support your revision, this Workbook contains two revision tasks to help you revise the skills that might be needed in your assessed task. Each task is divided into two sections:

1 Reading and making notes

In this Workbook you will use your skills to:
- Read and make notes on three provided research articles (pages 110–117 and 144–152).
- Carry out your own independent related reading to develop your understanding of the issues in the research articles and make notes (pages 118–125).

2 Answering questions

Your responses to the questions will help you to revise:
- Identifying and understanding **implications of key scientific issues** from the articles provided (pages 126–129 and 153–155).
- Understanding how organisations or individuals can **influence the scientific issues** (pages 130–132 and 156).
- **Assessing the validity of judgements made** (pages 133–136 and 157–159).
- **Suggesting implications** for **future developments** and/or **research** (pages 137–139 and 160).
- **Write your own article** based on a brief using appropriate terminology (pages 140–143 and 160–161).

> **Links** To help you revise skills that might be needed in your Unit 7 set task this Workbook contains two revision tasks starting on pages 109 and 144. See the introduction on page iii for more information on features included to help you revise.

108

Unit 7
Guided

Revision task 1

Revision task brief

You are provided with the following articles:

Article 1: Can I really stop malaria?

Article 2: Drug-resistant malaria an 'enormous threat' – vigorous international effort needed to contain it

Article 3: The ongoing battle against drug-resistant malaria

You need to become familiar with the articles and gain an understanding of the scientific issue discussed in the articles so that you are able to interpret, analyse and evaluate the articles.

Reading and making notes

> When reading and evaluating the articles provided, you need to break this into stages. Some suggested stages are given below. You will have a limited time in which to read, make notes and carry out any additional reading. Plan your time carefully to ensure you have completed everything within the allocated time. You should consider the following areas in your planning:
> - Understanding the scientific issue or issues in the articles.
> - The organisations/individuals mentioned in the articles and their potential influence on the scientific issue.
> - Considering validity of judgements in one or more of the articles.
> - Possible future areas for development or research linked to the main scientific issue.
> - Your understanding of the scientific issues and ability to explain them to a given audience.

> **Guided**
>
> Look at the sample plan below and use this to create your own plan of how to approach your reading related to the articles given below. You could check with your tutor or on the Pearson website how long you will have for your reading around the articles and include your own timings for each activity.

Task	
1	Reading the articles several times so they feel familiar.
2	Annotating the articles, making notes and identifying the science issues clearly.
3	Assessing the reliability of each article by finding out more about each publication/website, noting the audience for each.
4	Picking out any scientists or institutions highlighted in the articles. Looking them up to find out more, to help assess the reliability and status of both the scientists and the institutions.
5	Analysing any data presented in the articles, looking at sample sizes, length of study, funding, etc. If time, look up any original sources given for the data.
6	Looking up secondary sources – using keywords from the articles to help find other papers, articles or websites on the same science issues. Consider whether secondary sources largely support or refute the ideas in the set articles, assess their reliability and note them down with quotes you can use in the exam if you want to.
7	Making a table to compare and contrast the articles.
8	Organising notes and making sure annotations and references on the articles are clear and make sense.

Unit 7
Guided

> Read these three articles carefully – and read them more than once! As you read, think about the scientific issue involved. Look out for references to scientists, journals and institutions. Find connections between the three different articles. Then you'll be ready to annotate the articles and make extra pages of notes.

Article 1

Can I really stop malaria?
Blog by Annie Hillick, a concerned science writer

I don't know about you, but I find the news really depressing. There is so much awful stuff out there – and so little I can do about most of it. It makes me feel really helpless! So when I came across an amazing story the other day, where I could see light at the end of the tunnel and felt 'YES – I can really do something here!' – I thought I'd share …

It's about malaria. Malaria isn't a disease which affects us in the UK but, according to the United Nations, somewhere in the world a child dies every two minutes from this deadly disease. Nearly half of the population of the world is at risk from malaria, which is spread through a single bite from the female *Anopheles* mosquito. She needs a couple of protein-rich meals before she lays her eggs – and human blood gives her just what she needs. Mosquitoes are nocturnal feeders, so they bite and feed while people are asleep. Unfortunately, *Anopheles* mosquitoes don't just drink a tiny amount of blood. They also carry a blood parasite called *Plasmodium*. So, when an infected mosquito bites and feeds, she spreads the parasite into another victim.

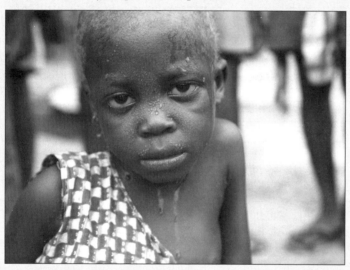

In 2015, there were 212 MILLION new cases of malaria and 429,000 deaths as a result of the disease. 70% of those who died were children. The great majority of those affected by malaria live in Africa but it also threatens people in South-East Asia, Latin America and the Middle East. Pregnant women, children, refugees and those affected by HIV/AIDS are most vulnerable to malaria – but no-one is immune to it. Malaria is a huge global problem – it has a devastating effect on individual people and their families – but it also condemns many communities and even whole countries to poverty, because people are often too ill to work effectively. So far, so depressing …

But here's the good bit: **MALARIA IS BOTH PREVENTABLE AND TREATABLE!**

Browsing the internet for resources when writing about infectious diseases, I came across a website called **NothingButNets.** It's a charity linked to the United Nations, working with a huge range of people, including the Bill and Melinda Gates Foundation. Their aim is, quite simply, to get rid of malaria – completely! It's taken us a long time to understand what causes malaria and how it is spread. But now we know, we can tackle this terrible disease. It seems possible, with enough investment and action from people around the world, that this global killer could eventually be eradicated completely. So, what can we do about it?

Treating malaria

Four Nobel Prizes have been awarded for work associated with malaria: to Sir Ronald Ross (1902), Charles Louis Alphonse Laveran (1907), Julius Wagner-Jauregg (1927), and Paul Hermann Müller (1948). Building on the discoveries of these great scientists we have produced some very effective antimalarial drugs – although oddly enough, two of the most effective treatments come from plants which have been known to have medicinal properties for centuries: artemisinin from the Qinghao plant (China, 4th century) and quinine from the cinchona tree (South America, 17th century).

However, the medicines which work well can be expensive and they can become ineffective over time. What's more, many of the people affected by malaria live in countries where there are few doctors and little health care infrastructure. That makes using medicines which need to be taken regularly over time really difficult. Artemisinin-based combination therapies (ACTs) can effectively cure malaria if they are started early enough. But like other medicines, ACTs are often not available in the remote areas of the world where they are most needed. What's more, it's proved really hard to make vaccines against malaria. Although some are now being trialled, they too have the disadvantages of being relatively expensive and hard to get to the people who need them. The simplest and best way to stop the devastation caused by malaria is to prevent it happening in the first place.

Preventing malaria

If malaria is spread by mosquitoes, the solution might seem obvious. Keep the people away from the mosquitoes. But in the poorest, and hottest, parts of the world, where malaria is at its worst, many people do not have windows or window screens, and doors are open to keep the inside of homes cool. Mosquitoes get in during the day, and at dawn and dusk. Then they become active and feed during the night, when people are asleep.

There is a really simple answer to this problem: long-lasting insecticidal bed nets (LLINs). These are mosquito-proof nets which are hung over the bed and tucked under the mattress. They have a powerful insecticide woven into their fabric. This gives a double whammy of protection – they provide a barrier between the mosquitoes and the sleeping people, and they kill any mosquitoes which land on them, reducing the mosquito population and so protecting the whole community. LLINs are cheap ($10 each), easy to use, and each net lasts at least 3 years. They are easy to distribute and very easy to use. Bed nets can reduce malaria transmissions by around 90% when lots of people in a community use them – and a 2012 World Malaria Report produced by the World Health Organization (WHO) showed that 90% of the people given bed nets use them. In the year 2000, only about 2% of people in Africa slept under bed nets. Now around 53% of the population have this amazing protection, and malaria rates are falling …

So here is a story which potentially has a happy ending.

Go to https://nothingbutnets.net/about/and see for yourselves.

Article 2

Drug-resistant malaria an 'enormous threat' – vigorous international effort needed to contain it

13:42, 20 FEB 2015   UPDATED 15:11, 20 FEB 2015
BY OLIVIA SOLON

The strain is spreading from Myanmar and health workers have very few other drug options

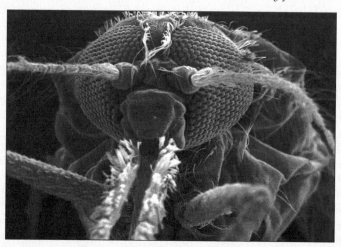

A drug-resistant form of malaria is spreading in southeast Asia and represents an "enormous threat" to the world.

The resilient form of the mosquito-spread parasite has been found in many parts of Myanmar (Burma) as well as Cambodia, Thailand, Laos and Vietnam.

The drug artemisinin is normally given as part of a combination therapy to battle the disease, but now a strain of the parasite is not responding to it.

The alarming discovery is a major blow to global health efforts to reduce the number of deaths from the mosquito-spread parasite.

Although initially other drugs given in the combination treatment could keep malaria at bay, the parasite is likely to develop resistance to the partner drugs as well – and there's evidence that this is happening already.

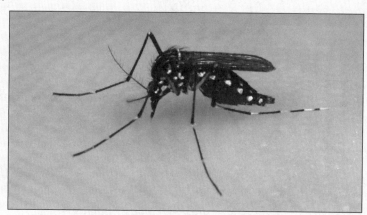

Malaria is spread by mosquitoes

Health workers fear that the strain could soon spread to India, where thousands more lives would be at risk.

"It's hugely worrying," says Professor David Conway from the London School of Hygiene & Tropical Medicine, "both for people in southeast Asia and the rest of the world."

"Should these [drugs] fail today, there's nothing waiting in the wings that's going to be affordable and adequately tested in time."

Something similar happened in the 1950s, when malaria became resistant to a drug called chloroquine. It spread across the world and eventually reached Africa.

"The global spread of chloroquine resistance resulted in the loss of millions of lives in Africa and, clearly, Myanmar is an important part of the frontline in the battle to contain artemisinin resistance," say the authors of the study, published in the *Lancet Infectious Diseases*.

"A vigorous international effort to contain this enormous threat is needed," they say.

Conway adds: "It's not too late, but action needs to be taken now to stop the spread."

Healthcare workers should make sure that malaria isn't being treated with artemisinin on its own (which is happening in some places "against all advice"), and efforts should be made to eliminate counterfeit medicines.

Continued molecular testing of the strain in real-time and monitoring patients in other territories, particularly Africa, for resistance is vital.

[Source: www.mirror.co.uk]

Article 3

The ongoing battle against drug-resistant malaria

Resistance to antimalarial drugs is one of the biggest problems currently facing malaria control. Recent studies looking at the genome of the malaria parasite could help scientists understand how drug resistance has evolved – and develop the tools needed to keep it in check.

Malaria occurs in more than 90 countries worldwide. On average it kills one child every minute and around half a million people every year. As a result, efforts to control the disease are a major global health priority.

Malaria-awareness publicity

Since 2000, funding for malaria control has increased and huge progress has been made, including improved management of the mosquito vector and the deployment of effective antimalarial drugs that work by killing the malaria parasite. Consequently, death rates have fallen globally by 47 per cent. However, the development and spread of drug-resistant malaria parasites is proving to be one of the greatest challenges to malaria control today today. *Plasmodium falciparum*, the most deadly species of malaria parasite, has developed resistance to nearly all antimalarial drugs currently in use.

What is drug resistance?

Drug resistance is the reduction in the effectiveness of a drug that has been designed to kill or inhibit a particular pathogen. It can arise as the result of one or more mutations in the genome of the pathogen that give it the advantage of being able to evade the effects of a drug.

Often drug resistance emerges gradually, starting initially with a handful of drug-resistant pathogens that survive exposure to a drug while all the drug-sensitive pathogens die. Having virtually nothing to stop them, neither drugs nor competition from drug-sensitive pathogens, the drug-resistant pathogens can multiply and their population grows.

How does drug resistance come about?

Malaria parasites are genetically very diverse and their genomes are changing (mutating) all the time. Occasionally, a genetic change can be beneficial, for example, by helping the parasite to hide from our immune system or by making the parasite resistant to a particular drug.

When inside the mosquito, the male and female forms of the parasite can pass these mutations to the next generation of parasites when they fuse and multiply. The resistant parasites will then enter another human when the mosquito bites and injects the parasites into their blood.

When working effectively, antimalarial drugs can clear all the malaria parasites from an individual's body in a matter of days. However, when drug-resistant parasites are present, the drugs either have no effect or are very slow to work, potentially leaving some parasites behind. This is called delayed parasite clearance. To rid an individual of malaria, every single parasite needs to be removed from the body, if not, malaria could more easily reoccur.

Looking back – the history of drug resistance

Malaria parasites have affected humans for thousands of years. During that time, humans have evolved mechanisms to protect themselves from the parasites and the parasites have evolved means of avoiding these defences.

However, with the introduction of the first drugs to tackle malaria, a new evolutionary battle was started: the parasites' struggle to evade antimalarial drugs. The first recorded use of an antimalarial drug was quinine in 1632 and, although it took the parasites a while (almost 300 years!), quinine-resistant parasites eventually emerged in 1910. Since then there have been several major drugs developed, but each time the malaria parasites have developed resistance:

- The drug chloroquine was introduced in 1945 with resistant parasites cropping up 12 years later.
- Sulfadoxine pyrimethamine was used from 1967 and resistant parasites were found in the same year!
- Mefloquine was given from 1977 but resistance was first recorded in 1982.

Scientists, doctors and patients have now been battling with drug-resistant malaria for over a century.

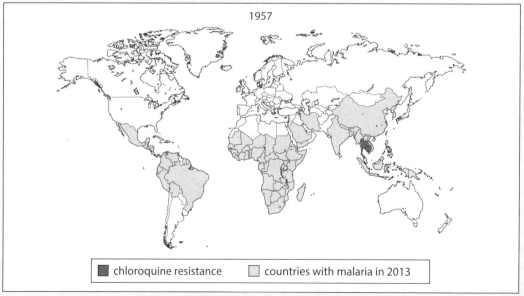

The spread of chloroquine resistance in countries with Plasmodium falciparum *from 1957 to 2005 (Data sources: Worldwide Antimalarial Resistance Network; WHO World Malaria Report, 2014). Image credit: Genome Research Limited*

Drug resistance déjà vu

Interestingly, drug resistance often emerges in the same place: the Greater Mekong Subregion in South East Asia.

Although it isn't known for sure, there are several theories as to why resistance may emerge in this region. One theory is that, compared to Africa, malaria infections are much less common in South East Asia. This means that drug-resistant malaria parasites face much less competition from other malaria parasites, and are therefore more likely to survive and reproduce.

Another theory is that because many people in Africa develop natural immunity to malaria, only a small fraction of all the people infected are actually treated with antimalarial drugs. This translates into a much lower drug pressure on the parasite when compared to South East Asia where antimalarial drugs are more widely distributed.

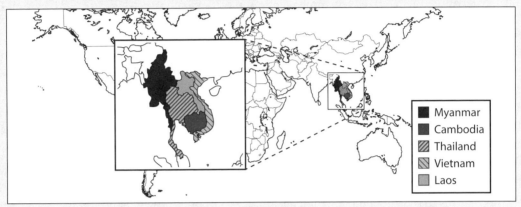

Map highlighting the Greater Mekong Subregion in South East Asia
Image credit: Genome Research Limited

Survival of the fittest

Perhaps surprisingly, for a malaria parasite a drug resistance mutation isn't necessarily very good for their biological fitness. Imagine all the drug-resistant parasites are wearing a big, heavy suit of armour. They are protected from the effects of the drug, but in terms of speed and agility they are less able to compete for food with parasites that are not drug-resistant. This means that they can only thrive in the presence of an antimalarial drug when the fast, but drug-sensitive parasites will be killed. We call it drug resistance; the malaria parasite calls it surviving!

Now there are only a limited number of drugs which can be used against malaria. The frontline treatment is currently artemisinin-based combination therapy.

Artemisinin-based combination therapy

Artemisinin-based combination therapy (ACT) has been integral to the recent successes in global malaria control. The main idea behind ACT was to provide an inexpensive, short-course treatment that would also help protect against the development of drug resistance. On paper, ACT should work perfectly. Artemisinin is a very fast acting drug which means that within 12 hours of starting treatment around half of the parasites in the body are removed. Artemisinin is combined with a partner drug that usually works more slowly, hammering the remaining malaria parasites until they are all dead.

One example of an effective ACT partner drug is piperaquine. First synthesised in the 1960s piperaquine was originally used alone as a malaria preventative and treatment in China. However, with the emergence of piperaquine-resistant *Plasmodium falciparum* in the 1980s, its use declined until it was deemed suitable for use in combination with artemisinin. This is due to its low cost, high efficacy and minimal side effects. A partner drug, like piperaquine, provides a counter punch so that if malaria parasites develop resistance to artemisinin, they will still be knocked out by the partner drug.

Is history repeating itself?

In 2009, researchers reported concerns that artemisinin was taking longer to clear parasites from patients infected with *Plasmodium falciparum* along the Thailand–Cambodia border — a worrying sign of emerging drug resistance. Since then, researchers have reported slow parasite clearance in four countries in the Greater Mekong Subregion and in some locations they've even seen treatment failures. This puts additional pressure on the partner drugs to kill the parasites, which may lead to resistance to these drugs too.

What's more, if artemisinin resistance were to arise in Africa or emerge independently elsewhere, as has happened with other antimalarial drugs, the public health consequences would be catastrophic. It is likely there would be a reversal in the recent declines seen in malaria mortality rates and the number of deaths due to malaria would start to increase.

In response to this threat, the World Health Organization (WHO) launched an emergency plan of action to tackle artemisinin resistance in the Greater Mekong Subregion covering 2013–2015. They proposed an immediate and coordinated increase in efforts to tackle malaria in Cambodia, Laos, Myanmar, Thailand and Vietnam.

Currently the WHO's goal is to initiate elimination activities by 2020 in order to remove malaria completely from Greater Mekong Subregion countries by 2030.

But how are we going to stop the spread of drug resistance if we haven't been able to in the past? Well, now we have one more weapon in our arsenal that we didn't have before – genome sequencing!

Genomics vs. malaria – the fight is on

To develop an effective strategy to combat malaria once and for all it is crucial to understand the genetic factors that determine how drug resistance emerges and spreads.

At the time that artemisinin resistance was first discovered in early 2009, no one knew which genetic changes were responsible, and pinpointing those changes proved more challenging than expected.

However, faster and cheaper genome sequencing techniques have enabled us to learn a lot more about the underlying genetic changes responsible. Scientists have now compared thousands of parasite genomes from different areas of Africa and South East Asia to identify the genetic variations that could lead to drug resistance. By finding these genetic changes scientists are hoping that they may eventually be able to track and then prevent the spread of artemisinin resistance.

Clues on chromosome 13

In 2012, and then again in 2013, a couple of genome-wide association studies (GWAS) looking at the *P. falciparum* genome pointed towards two regions next to each other on chromosome 13 as potential sites of the mutations associated with artemisinin resistance. However, they needed to find out for sure if these mutations were directly involved in resistance. A year or so later, a collaboration led by scientists at the Institut Pasteur in Paris came up with an experiment that pointed them in the right direction.

Over a five year period, the scientists grew and nurtured a strain of *Plasmodium* parasite that they knew did not have any resistance to artemisinin. Every so often, during this period they gave the colony of parasites a small amount of artemisinin. They hypothesised that sooner or later an artemisinin-resistant parasite would emerge because of the selection pressure of the drug (the pressure to adapt in order to survive!). Sure enough, after four years of exposure to the drug, artemisinin-resistant parasites were seen. With DNA sequencing they were then able to study the genome of the resistant parasites and compare them to the genome of the original, non-resistant strain of *Plasmodium*.

They found several genetic changes in the resistant parasite genome but the most significant one occurred bang in the middle of the previously-identified regions on chromosome 13, in a gene called *kelch13*.

What does *kelch13* do?

kelch13 is one of the most conserved genes in the *Plasmodium* genome. This means that it has remained pretty much the same for about 50 million years. This suggests that its function must be crucial for the survival of the *Plasmodium* parasite. As a result, although a mutation in this gene may make it resistant to artemisinin, it would also probably reduce the normal biological fitness of the malaria parasite, as if they had put on a heavy suit of armour.

Searching across the genome

To better understand how malaria parasites are developing drug resistance in the wild, researchers carried out a study of *P. falciparum* parasites sampled from more than 1,000 malaria patients in South East Asia and Africa. The researchers collected blood samples from patients before artemisinin-based combination therapy (ACT) was given to them and then every six hours during treatment with ACT to track how quickly the parasites were being cleared from the body.

They then isolated parasite DNA from all of the blood samples and determined the parasites' genome sequence. By comparing a large number of parasite genomes it enabled the researchers to study genetic differences between drug-sensitive and drug-resistant parasites. The results were quite unexpected.

In the drug-resistant parasites they didn't just find one mutation in *kelch13*, they found more than 20! This explains why it was so hard to pinpoint the exact mutation. There wasn't a single mutation that caused artemisinin resistance but several mutations, each emerging at different times and places in the *kelch13* gene.

Kelch13 and friends

When scientists looked beyond *kelch13* at the rest of the genome there was another surprise in store. They discovered mutations in at least four other genes (*fd*, *arps10*, *mdr2* and *crt*) that are also associated with artemisinin resistance. Whenever the *kelch13* mutation was present in the genome of a resistant parasite, the other four mutations almost invariably seemed to be there too. Was this just a coincidence?

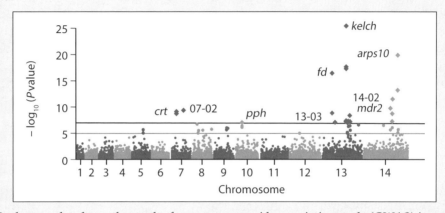

This Manhattan plot shows the results from a genome-wide association study (GWAS) investigating artemisinin resistance. Each dot represents a single letter change in the parasite genome and its location on a chromosome. The vertical height of the dot on the plot shows how strongly associated that genetic change is to artemisinin resistance. You can see a mutation in the kelch gene on chromosome 13 is most strongly associated with artemisinin resistance. You can also see that mutations in other genes, arps10, fd, mdr2 and crt, are also associated with artemisinin resistance.
Image credit: Genetic architecture of artemisinin-resistant Plasmodium falciparum, *Nature Genetics 47, 226–234 (2015) doi:10.1038/ng.3189.*

Although it is not yet known what the exact role of the four mutations might be in drug resistance, it might be that they provide an ideal environment for the *kelch13* mutation to arise. They may, for example, compensate for the parasite's reduced fitness due to artemisinin resistance. If we think back to drug resistance being like a heavy suit of armour, these four background mutations could be providing the resistant parasites with a horse that helps them maintain their mobility and compete more equally with susceptible parasites for food.

Whatever their role might be, discovering these mutations has given scientists a useful tool for monitoring the spread of artemisinin resistance. Finding this fixed set of four mutations in parasite populations could act as a warning sign that these parasites are at high-risk of developing artemisinin resistance. This will enable researchers to target those areas with medicines or insecticides to help kill any artemisinin-resistant parasites before they take hold and spread any further.

Looking forward

Any intervention we carry out to control malaria has an impact. We can think of malaria control interventions as large-scale evolutionary experiments. By introducing a drug or a vaccine against a parasite, we are applying a selection pressure and encouraging those genetic changes that will enable the parasite to survive in the presence of a drug or vaccine.

Genomics is a powerful tool to help us observe the evolutionary impact of these interventions. We can then use this information to inform decisions about which methods to use in future. For example, if the malaria parasite starts to show signs of resistance to one drug, it is possible to switch to another drug or change the drug regimen.

Like spies in an enemy country, genomics can provide us with the intelligence to track drug resistance emerging in the malaria parasite. This gives us more time to plan our counterattack before drug resistance becomes more widespread.

[Source: http://www.yourgenome.org]

Unit 7
Guided

> **Guided** **Familiarise yourself with Article 1**
>
> It's worth reading each article at least twice before you start to annotate or highlight the text.
>
> Think carefully – make sure you really grasp what the article is about before you start writing yourself.
>
> Here is part of the first article with some annotations by a student. Using different colours to highlight or underline different features helps identify the key points. Have a look – can you think of anything else to add?

… according to the United Nations, somewhere in the world a child dies every two minutes from this <u>deadly disease</u>. Nearly half of the population of the world is at risk from malaria, which is spread through a single bite from the female *Anopheles* mosquito … Unfortunately, *Anopheles* mosquitoes don't just drink a tiny amount of blood. They also carry a blood parasite called *Plasmodium* … In 2015, there were <u>212 MILLION new cases of malaria and 429,000 deaths</u> as a result of the disease. 70% of those who died were children.

> Scientific issues – malaria as a major killer disease.

> Highlighting or underlining text which emphasises the scientific issues like this is a good idea.

… two of the most effective treatments come from plants which have been known to have medicinal properties for centuries: <u>artemisinin from the Qinghao plant (China, 4th century) and quinine from the cinchona tree (South America, 17th century)</u> …

However, the <u>medicines which work well can be expensive</u> and they can become ineffective over time. What's more, many of the people affected by malaria live in countries where there are <u>few doctors and little health care infrastructure</u>. That makes <u>using medicines which need to be taken regularly over time really difficult</u>. Artemisinin-based combination therapies (ACTs) can effectively cure malaria if they are started early enough …

If malaria is spread by mosquitoes, the solution might seem obvious. Keep the people away from the mosquitoes. But in the <u>poorest, and hottest, parts of the world</u>, where malaria is at its worst, <u>many people do not have windows or window screens, and doors are open</u> to keep the inside of homes cool. Mosquitoes get in during the day, and at dawn and dusk. Then they become active and feed during the night, when people are asleep.

Bed nets can reduce malaria transmissions by around 90% when lots of people in a community use them – and a <u>2012 World Malaria Report produced by the World Health Organization</u> (WHO) showed that 90% of the people given bed nets use them.

> Future research – Finding new drugs against malaria will be important **future research**.

> Scientific issues – problems of treating malaria in the countries it affects.

> Look up any reports mentioned in the text and the organisation that produced them, such as this report by the WHO, to see if they are reliable evidence.

> **Guided** Go back to Article 1, starting on page 110, and highlight and annotate the pages, using these examples to get you started.

118

Unit 7
Guided

> **Guided** **Familiarise yourself with Article 2** Once you have read all the articles thoroughly and annotated Article 1, go through Article 2 in the same way. Highlight important points, and make annotations in the margins to help you. Here is part of the second article, also with some annotations by a student. Have a look – can you think of anything else to add?

A drug-resistant form of malaria is spreading in southeast Asia and represents an "enormous threat" to the world.

<u>The resilient form of the mosquito-spread parasite has been found in many parts of Myanmar (Burma) as well as Cambodia, Thailand, Laos and Vietnam.</u>

The drug artemisinin is normally given as part of a combination therapy to battle the disease, but now a strain of the parasite is not responding to it.

> *Scientific issue - treating malaria and problem of drug-resistant disease-causing parasites.*

> *Highlighting the scientific issues across all three articles helps you see the common content.*

The alarming discovery is a major blow to global health efforts to reduce the number of deaths from the mosquito-spread parasite …

"It's hugely worrying," says <u>Professor David Conway</u> from <u>the London School of Hygiene & Tropical Medicine</u>, "both for people in southeast Asia and the rest of the world."

> *London School of Hygiene and Tropical Medicine is a world-leading centre for research and post-graduate education in public and global health (www.london.ac.uk). Ranked third in world for social sciences and public health and top in Europe for research impact (Wikipedia)*

> *Underline or highlight the names of individual scientists and scientific institutions in the articles. Look them up and make brief notes about them. This can tell you how well respected and influential they are. Always note down your sources!*

"Should these [drugs] fail today, there's nothing waiting in the wings that's going to be affordable and adequately tested in time." …

<u>"The global spread of chloroquine resistance resulted in the loss of millions of lives in Africa and, clearly, Myanmar is an important part of the frontline in the battle to contain artemisinin resistance,"</u> say the authors of the study, published in <u>the *Lancet Infectious Diseases* …</u>

Continued molecular testing of the strain in real-time and monitoring patients in other territories, particularly Africa, for resistance is vital.

[<u>Source</u>: www.mirror.co.uk]

> *Reinforces idea that new drugs against malaria might be important future research.*

> *The Lancet is one of the oldest and best known medical journals in the world. Weekly and peer-reviewed.*

> *New, quick, reliable tests might be research for future?*

> *The Mirror is a popular newspaper BUT the journalists have based this report on well-respected scientific publications such as the Lancet*

> *Comment on the source of the article and likely reliability.*

> **Guided** Go back to Article 2, starting on page 112, and highlight and annotate the pages, using these examples to get you started.

Unit 7
Guided

> **Guided** **Familiarise yourself with Article 3** In this example, the third article is much longer than the others. Here are some excerpts from the article with annotations a student has made. How many points do you think they have missed?

Resistance to antimalarial drugs is one of the biggest problems currently facing malaria control. Recent studies looking at the genome of the malaria parasite could help scientists understand how drug resistance has evolved – and develop the tools needed to keep it in check.

Confirms scientific issue – the problems of treating malaria and drug-resistant malaria parasites.

Malaria occurs in more than 90 countries worldwide. On average it kills one child every minute and around half a million people every year. As a result, efforts to control the disease are a major global health priority …

To better understand how malaria parasites are developing drug resistance in the wild, researchers carried out a study of *P. falciparum* parasites sampled from more than 1,000 malaria patients in South East Asia and Africa …

1000 patients = good sized study, fairly large sample.

1957

■ chloroquine resistance   □ countries with malaria in 2013

A map showing the spread of chloroquine resistance in countries with Plasmodium falciparum *from 1957 to 2005. (Data sources: Worldwide Antimalarial Resistance Network; WHO World Malaria Report, 2014). Image credit: Genome Research Limited*

What could be commented on here?

Source: www.yourgenome.org

Website by the Wellcome Trust genome campus – top international centre for DNA sequencing, one Nobel Prize winner (Wikipedia).

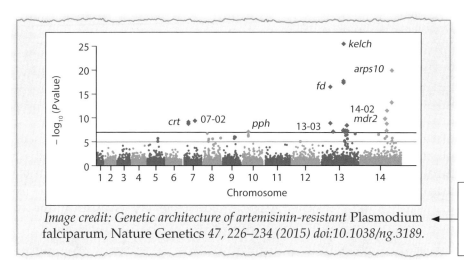

Image credit: Genetic architecture of artemisinin-resistant Plasmodium falciparum, *Nature Genetics 47, 226–234 (2015) doi:10.1038/ng.3189.*

Peer-reviewed journal publishes highest quality research in genetics (Nature.com; Wikipedia).

> **Guided** — Go back to Article 3, starting on page 113, and highlight and annotate the pages, using these examples to get you started.

Unit 7
Guided

Making notes

You may be allowed to take some of your preparatory notes into your supervised assessment time. If so, there may be restrictions on the length and type of notes that are allowed. Check with your tutor or look at the latest Sample Assessment Material on the Pearson website for more information.

 Guided

It will be helpful to ask yourself some key questions when making notes about the articles. First of all identify **the main scientific issue** covered by the three articles. Having read the articles, you should have a good idea of what the main scientific issue is but there may also be related issues. When you re-read the articles, highlight anything about the scientific issues covered so you can quickly and easily refer to the relevant sections in your notes. Complete the sample student notes below on the key issues covered.

The main scientific issue is the problem of malaria and the development of drug-resistant strains of the parasite which causes the disease, e.g. in Article 1: 'In 2015, there were 212 MILLION new cases of malaria and 429,000 deaths as a result of the disease' – most of these were children under the age of 5.

In Article 2: A drug-resistant form of malaria is spreading in South East Asia and represents an 'enormous threat' to the world. In Article 3: Resistance to antimalarial drugs is one of the biggest problems currently facing malaria control.

The secondary issues include how do we treat or prevent malaria and how can we deal with the problem of the drug-resistant parasites. Ideas from the articles include:

....

....

....

 Guided

Next, look for key words in the articles you are given. These will help you **find supporting evidence on the scientific issues** highlighted in your articles. You might find this supporting evidence on the internet. Magazines like *New Scientist* or scientific journals in your library can also be good sources of supporting evidence. Always note down the sources of this supporting evidence. Complete the student notes below on key words and supporting evidence in the three articles given.

Key words: malaria; drug-resistant malaria parasites; treating malaria;

....

Supporting evidence: World Health Organization at http://www.who.int/malaria/areas/drug_resistance/overview/en/ This link takes you to the WHO website. This reliable source confirms that parasite resistance to antimalarial drugs has been documented in three of the five malaria species known to affect humans. It recommends artemisinin-based combination therapies (ACTs) for the treatment of uncomplicated malaria. The WHO states clearly that resistance of the malaria parasite is an urgent

public health concern which could threaten efforts to control malaria all over the world ....

....

http://www.nature.com/scitable/knowledge/library/evolution-of-drug-resistance-in-malaria-parasite-

96645809 The evidence at this URL, from the reputable journal *Nature*, tells us that ....

....

....

Pick out any individual **scientists** named in the articles. Identify any **scientific organisations** mentioned.

You can also identify any **named journals** where evidence is published.

Do some background research on each journal. Find out their status and consider how influential they might be. Note your sources.

Decide how you want to display the information. The student below has used a table in their notes to show the different scientists mentioned in the articles. They have used a list of bullets to outline the scientific organisations. Add some further information where it's required.

Article	Details about the named scientists
1	All won Nobel Prizes so good reputations and influential
	Sir Ronald Ross (1902) – a doctor who demonstrated that malaria is spread by mosquitoes (Wikipedia, Nobel Prize website)
	Charles Louis Alphonse Laveran (1907) ....
	....
	Julius Wagner-Jauregg (1927) – doctor who discovered a treatment for paralysis from syphilis by infecting patients with malaria to give them a high fever which cured the paralysis (Nobel Prize website, Wikipedia)
	Paul Hermann Müller (1948) ....
	....
2	Professor David Conway – professor at the London School of Hygiene and Tropical Medicine, works in UK and African countries, published over 170 research articles, well known in his field (LSHTM website)
3	No individual scientists mentioned

Article 1 organisations

- <u>World Health Organization (WHO):</u> WHO is a specialised part of the United Nations and is very influential – it collects data on diseases from all over the world and issues guidelines for treatments, flags up epidemics and pandemics, etc. It put together the 2012 World Malaria Report based on data from 104 countries with endemic malaria. VERY influential globally (WHO publications website, Wikipedia).

Article 2 organisations

- <u>London School of Hygiene and Tropical Medicine:</u> This is a world-leading centre for research into tropical medicine and public health. Part of the University of London and founded 1899. Research carried out here has <u>big impact all over the world</u> (2016 CWTS Leiden Ranking for research impact).

Article 3 organisations

- World Health Organization (WHO): see Article 1

....

....

....

Unit 7
Guided

> Your notes on the scientists and scientific institutions referred to in each article start to give an idea of their validity. Now make notes on **where the articles come from, and how reliable any data they publish seem to be**. You can make lists, or tables, or just jot down notes summarising your ideas. Some sample notes are given below to get you started.

<u>Article 1</u> refers to the 2012 Malaria Report from the WHO – a very reliable source. It is a blog written by a concerned science writer so there may be bias to information which highlights problems and makes the situation sound as bad as possible.

<u>Article 2</u> is from the website of the <u>Mirror</u> newspaper. This is not a heavyweight newspaper so reporting may not be valid, BUT the report is based on work by Professor David Conway in a study published in the <u>Lancet Infectious Diseases</u>, a very reputable scientific journal which only publishes peer-reviewed valid work.

<u>Article 3</u> from www.yourgenome website produced by Wellcome Trust Genome Campus ....

....

....

The studies listed in the data are ....

....

....

Data published in <u>Nature Genetics</u>, ...

....

....

> **Read around the scientific issues.** Note down your reading and any useful quotes which support or disagree with the articles you have been given. At the same time make notes on **possible future areas for development or research** linked to the main scientific issue.

<u>Supporting evidence:</u>

1. <u>New Scientist</u> 2015 'Drug-resistant malaria poised to cross into India': confirms development of drug resistance in Cambodia, mutations in K13, references Institut Pasteur

2. ....
   ....
   ....

3. ....
   ....
   ....

4. <u>Independent:</u> British armed forces set to ban most prescriptions of controversial antimalarial drug Lariam ... Antimalarial drug very effective against parasite but can cause mental health problems
   ....
   ....

Future developments:

- new drugs to cure malaria
- genome analysis to find out where most resistance is found
- ....
- ....
- ....

> **Guided**
>
> **Comparing the articles** is a great way to summarise all the differences and similarities you have discovered. There are lots of ways to do this but a table is often the most clear and easy to use. Complete this table in your own words.

	Article 1	Article 2	Article 3
Source	blog	http://www.mirror.co.uk/news/technology-science/science/drug-resistant-malaria-enormous-threat---5197706 Popular newspaper	http://www.yourgenome.org/stories/the-ongoing-battle-against-drug-resistant-malaria Website produced by Wellcome Trust genome Campus – top genome research institution in the world
Scientific issues	Global problem of malaria and treatments		Global problem of malaria and drug-resistant malaria parasites
Individuals mentioned and influence	Four Nobel laureates: Sir Ronald Ross (1902), Charles Louis Alphonse Laveran (1907), Julius Wagner-Jauregg (1927), Paul Hermann Müller (1948) Huge influence, but not very important in article		
Organisations mentioned and influence		London School of Hygiene and Tropical Medicine – highly influential	World Health Organization – highly influential
Sources			Nature Genetics
Validity		Valid	
Future development		Need to deal with drug resistance Need for new drugs	

Answering questions

Here are some examples of skills involved in analysing the articles provided and answering questions based on them.

Answer all the questions in the spaces provided

1 Discuss the implications of the scientific issue identified in the articles.

> Look back at your notes to remind yourself about the main issue in the three articles. Make sure you refer to the articles and to other reading you may have done. Your answer will need to be clear, coherent and logical, so it is worth spending time planning your answer. Read the sample student plan below and then use it to complete the response that comes after it.

<u>Plan for my answer</u>

1. Outline main issue: spread of drug-resistant malaria
2. Explain what is malaria, who it affects and where
3. Outline possible treatments (insecticides, vaccines, drugs, nets)
4. Discuss issues in treating malaria (cost, environmental issues, ineffective drugs and resistance, difficulty getting drugs/vaccines to right people, difficulties in getting people to use nets)
5. Focus on drug-resistant malaria (background, problems (Article 3), genome analysis (Article 3), costs (Article 2), difficulty getting people to take drugs (Article 2)
6. How to tackle drug resistance
7. Global implications of drug-resistant malaria

> You need to draw a wide range of links to and between the ethical, social, economic and/or environmental implications of the science.
>
> Make sure you refer to each of the different articles in answering this question.

The main scientific issue identified in these three articles is the spread of drug-resistant malaria in many parts of the world. There are two aspects to this issue. One is the problem of malaria as a global disease. The other is the problem of treating malaria successfully, especially as forms of the malaria parasite are evolving which are resistant to our best antimalarial drugs.

> It's a good idea to start out by stating the big scientific issue.

Malaria is a global disease which affects ....
....
....
....
....

> Carry on here by explaining why malaria is such a problem. Describe what it is, how many people it affects and the social and economic problems this causes.

It isn't easy to treat malaria. Many of the countries where malaria is a big problem are very poor. The drugs used to treat the disease can be expensive. So we try to prevent the disease from spreading in several different ways.

....

....

....

....

....

....

> Continue this by explaining the impact of different control methods. For example, you might mention the environmental problems caused by using insecticide sprays on rivers to kill the larvae; social problems of getting people to use mosquito nets; or economic problems with the expense of insecticide spraying, medicines and vaccines.

> Make sure you mention low-tech solutions such as insecticide-impregnated mosquito nets, and the pros and cons of such approaches.

....

....

....

....

....

....

....

....

....

It has been possible for many years to treat malaria successfully but only if the treatment is started early. Quinine was discovered in the 17th century and artemisinin was first discovered even earlier. But there are lots of issues and implications when we try to cure malaria with drugs or vaccinations.

> Here the student has explained that treatment needs to start early to be effective and has mentioned two of the key antimalarial drugs. Use your notes and the articles to help you write more about some of the different drug treatments that are used against malaria, especially the ACTs.

These include ....

....

....

....

....

....

....

....

127

Unit 7
Guided

> Next you should give some of the reasons why drug treatments for malaria can cause problems of different sorts. You can comment on the cost of the drugs (economic and ethical issues), the problems of getting people in communities in affected countries to take the drugs regularly (social issues) and the risk of side effects from the drugs (ethical issues). In every case, support your statements with references either to the articles or to further reading you have done.

Although malaria treatments can be very effective, there are many different issues. The areas of the world where malaria is the biggest problem are often relatively poor countries with low levels of education in the population – for example Article 2 highlights sub-Saharan Africa as an area with big malaria problems. The people in these areas often cannot afford the drugs and it can also be difficult to get the drugs to the people. These are important economic and ethical considerations. Without education and an understanding of how malaria is spread, people do not always recognise the importance of taking antimalarial drugs correctly – so drugs that need to be taken for a long time are no use. Article 2 highlights the need for healthcare workers to help make sure drugs are used

correctly so they are as effective as possible. ....

....

....

....

....

....

One of the biggest scientific issues linked to malaria is the development of drug resistance in the parasites which cause the disease. The implications of this are huge, both for the countries affected and countries which do not yet have malaria.

> Explain clearly what is meant by drug resistance, what problems this causes and the implications for everyone. Make sure you give clear examples from the articles provided.

Drug resistance happens when a mutation takes place in the malaria parasites so they are no longer

affected by a particular antimalarial drug. For example Article 3 explains ....

....

....

....

....

....

....

Once the malarial parasites become resistant to one medicine there are big problems for the people who have malaria, the doctors who treat them and the scientists trying to get rid of the disease. One important way of tackling drug resistance has been to combine the treatments.

> Describe combination therapy and how it has helped. Then mention the development of parasites resistant to the combined therapies.

....The global implications of drug-resistant malaria parasites include the fear that one day we will not be able to treat the disease. This has enormous financial implications for the countries where the disease is common. It means millions of their children will die. It also means that many of the people who would normally be working and earning money to support their families and the economy of the country cannot work because they are too ill. The cost of

looking after so many ill people ....

> Highlight the social, economic and environmental impacts of uncontrolled malaria.

With global warming there are implications for areas where malaria is not yet a major problem.

> Explain the potential environmental effects of global warming on the distribution of the mosquitoes which carry the malaria parasites. Then discuss how this might affect the health of people in areas such as Europe.

Unit 7
Guided

2 Identify the different organisations/individuals mentioned in the articles and suggest how they may have had an influence on the scientific issue.

> Look back at your notes to remind yourself about the organisations and individuals mentioned in the articles.
>
> If you have made good notes, you will have these organisations and individuals already identified which will make answering this type of question much easier.
>
> You need to make clear links between the institutions and people mentioned and the original articles. Use your notes to complete the sample student response below.

There are three major organisations mentioned in these articles. The World Health Organization (WHO) is mentioned in Articles 1 and 3, the London School of Hygiene and Tropical Medicine is mentioned in Article 2 and the Institut Pasteur is mentioned in Article 3 for its work on the genomes of malarial parasites.

The World Health Organization, known as the WHO, is part of the United Nations.

> Use your notes to help you explain what the WHO is and why it is so influential.

....
....
....
....
....

The London School of Hygiene and Tropical Medicine is mentioned in Article 2. It is a world-leading centre for research into tropical medicine and public health. It is part of the University of London and it was founded in 1899. It specialises in tropical diseases such as malaria and specialist scientists come here from many different countries to study these diseases and how to treat them. Research carried out here has big impact all over the world.

> Use the research you did when you made your notes as you explain the influence of the Institut Pasteur.

The Institut Pasteur is in Paris. ....
....
....
....
....
....
....
....
....

All of the people mentioned in the articles are scientists who work in well-known institutions and are influential. They have influenced our understanding of how malaria is transmitted and how it can be treated — some of them are still influential. They include four Nobel Prize winners and Professor David Conway, a scientist who is still active and working in the very influential London School of Hygiene and Tropical Medicine.

> Here is another opportunity to use the notes you have made to give the names of the Nobel Prize winning scientists mentioned in the articles, along with a brief mention of what they did to influence what we now know about malaria and how to defeat it.

The four Nobel Prize winners are all mentioned in Article 1. They played a historical role in developing our understanding of how malaria is spread and how we might treat it.

....
....
....
....
....
....
....
....
....
....

Professor David Conway is very active in antimalaria research. He is raising awareness of the problems globally — in Article 3, by the hugely influential Wellcome Trust, he is quoted saying:

'… It's hugely worrying,' says Professor David Conway from the London School of Hygiene & Tropical Medicine, 'both for people in southeast Asia and the rest of the world.'

'Should these [drugs] fail today, there's nothing waiting in the wings that's going to be affordable and adequately tested in time.'

> Use your notes and references in the articles to give more details of Professor Conway's influence.

....
....
....
....
....
....
....
....
....

Unit 7
Guided

3 Discuss whether Article 3 has made valid judgements.

In your answer you should consider:
- how the article has interpreted and analysed the scientific information to support the conclusions/judgements being made
- the validity and reliability of data
- references to other sources of information.

> In order to answer this question well you will need to use your notes on Article 3 to help you highlight the main points. As you will be assessed on the structure of your discussion as well as the content, it must be clear, coherent and logical. This means it is worth taking time to create a plan for your answer. An answer plan could be in the form of a list, a spider diagram or another format. Complete the example spider diagram plan for this question.

Plan for my answer

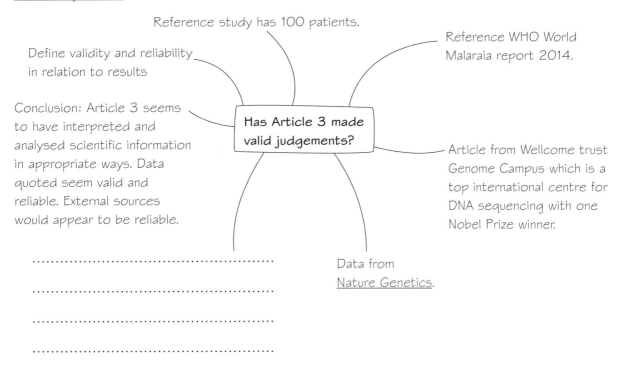

> Look back at your notes to remind yourself about the validity of the article. What does validity mean? What will you be looking for?
> Make sure you refer to the bullet points given in the question to guide your answer.

If results are **valid** they measure what they are supposed to measure. If results are **reliable**, the investigation produces stable, consistent results which other people can replicate. It is important to be sure that an article is valid and reliable before you take any notice of it. Here is a summary of the validity and reliability of the data used in these three articles, and how they have interpreted and used the data to support their conclusions. There is considerable agreement between the three articles, which in itself suggests that the conclusions are probably reliable.

Unit 7
Guided

> Use your notes to give the sources of the data used in Article 3. Comment on the likely validity of the data they have used.

Article 3 is much longer than the other two articles. It is published on the www.yourgenome.org website developed by the education team at the Wellcome Trust Genome Campus. This is the top institution globally for analysis of the genomes of different organisms, and it has many internationally renowned scientists working there on various projects. That alone suggests that the content will be both valid and reliable, and that any conclusions drawn will be supported by evidence. Reading through the article, evidence and data are presented from a variety of sources. ....

....

....

....

....

> In the first paragraphs the answer summarises the meaning of validity and reliability and looks at Article 3. They conclude that the data sources used in that article are valid and the conclusions drawn are reliable, as a result of the sources used and the fact that the conclusions are echoed in other sources.

> In this part of the answer the student is analysing Article 3. Use your own notes to complete this section of the answer, commenting on the sources of the data presented and the scientists involved in Article 3.

Unit 7
Guided

....
....
....
....
....
....
....
....

The sources quoted in this article are very reliable. The content also confirms the impact of malaria, some of the problems in preventing malaria and the problems of drug-resistant malaria parasites described in Articles 1 and 2. For example, Article 3 quotes vector management as well as effective antimalarial drugs as an important reason for why death rates from malaria have fallen by 47% since the year 2000. This reinforces the message in Article 1 about the importance of simple methods of vector control such as insecticide-impregnated mosquito nets. Later in Article 3, it confirms the conclusion from Article 2 that malaria parasites have become resistant to various drugs and that the current drug resistance has emerged from South East Asia.

> Give a historical perspective from Article 3, showing how malarial parasites have developed resistance to several drugs over the years, and that the resistance always seems to emerge in the same part of the world.

....
....
....
....
....

Article 3 gives us a lot more information on the genetic basis for development of drug resistance in malarial parasites. Considering where the article was developed, and the quoted sources, we can be fairly certain that the judgement made in the article that genome analysis of the malaria parasites is a vital tool in the ongoing battle against drug-resistant malaria is completely justified.

> Choose another couple of examples from Article 3 that convince you that the content is valid and reliable and confirms the conclusions of the other articles.

....
....
....
....
....
....
....

Unit 7
Guided

Unit 7
Guided

> **Guided** 4 Suggest potential areas for further development and/or research of the scientific issues from the three articles.

> Look back at your notes. When you were reading the articles and making notes you will have thought about potential future development and research. In this answer you can suggest more than one direction for future R&D. You will need to justify your choices by explaining why that research is needed. You are expected to link your ideas for further development and research to science from the three articles.

The big problems highlighted in these articles are the issues of malaria, the toll it takes on human lives and the economies of affected countries – and the growing problem of drug resistance in the malarial parasite which causes the disease.

Here are some potential areas for further research and/or development:

1 New drugs to cure malaria which work in a different way to the current medicines. This would mean that even the malarial parasites resistant to the current drugs would be wiped out as they would not have resistance to the new drug mechanism. A new drug like this could be used in combination with a current antimalarial drug so it continues to be effective. This would be similar to the use of piperaquine with artemisinin.

2 Genome analysis to find out where the drug resistance genes are found ....

> Complete with an explanation of how this line of new research might help solve the issues raised by malaria and drug-resistant malaria parasites, linking to genome analysis as explained in Article 3.

....
....
....
....
....
....

3 Development of new insecticides to be used on water or on mosquito nets ....

> Complete with an explanation of how development of new insecticides can help control the mosquitoes which spread malaria and so reduce the need for new antimalarial drugs.

....
....
....
....
....
....

Unit 7

Guided

> Add any further ideas for research and development of your own.

5 You are working with a charity which is raising funds to support research into new ways of treating drug-resistant malaria. You have been asked to write an article for a popular national newspaper. Your editor wants you to make as many people as possible aware of the problems of drug-resistant malaria, including the science issues.

Use information from the three articles provided for this task.

When writing your article, you should consider:
- who is likely to read the article
- what you would like the reader to learn from the article.

> You should plan your answer and make sure you consider the following points:
> - Keep your target audience in mind.
> - Plan what you want them to know.
> - Decide how you can include good, valid, reliable science but still make your article interesting – you want people to keep reading to the end.
> - A good headline will catch your reader's attention.
> - Use information from all the articles to show the examiners that you have read and taken in information from all your sources and understand where they agree and where they take different approaches.
> - If you can, include some extra information from your reading around.
> - Keep your tone, style and level of scientific terminology the same throughout the article.

> You should ensure that you write a well-organised article, with a clear, logical, coherent structure – so make your planning time count. Read the sample student plan for the article and then complete it below.

Plan for my article

Introduction: Statistics on malaria and who is affected. Number of deaths, number of people infected, cause and why should we care in the UK?

Drug resistance: How malaria is becoming more dangerous through drug resistance. Ways to control this and problems with these.

How is it being tackled? Describe use of nets, development of vaccines based on genome analysis.

Call to action: Explain our research and why more funding is needed.

> You can start a newspaper article with a headline which uses emotive terms like 'time-bomb' and makes it feel personal to the audience with the use of the emphasised 'YOU'.

A malarial time-bomb – and it could affect YOU!

Every year about 600 000 people die of malaria. Most of them are children under 5 years old. The World Health Organization estimates that round 219 million people are infected with malaria. It is caused by a tiny parasite, spread by the bites of infected mosquitoes. Malaria wrecks individual lives and destroys economies when many of the working population are infected with this dreadful disease.

So what? Malaria doesn't affect us here in the UK – but only because it is too cold for the mosquitoes which carry the malaria parasite to survive. In Shakespeare's times malaria (the ague) was common. As global warming increases, our old enemy could return with a vengeance.

This answer starts off with some shocking statistics about the effect malaria has on people around the world. Then it brings the problem closer to home, with the alarming idea that global warming could result in malaria even in the UK. This student has done some reading so they know malaria used to exist in this country.

Now describe why malaria is becoming even more dangerous than ever. You might begin by talking about ways in which we have managed to start controlling malaria – and then introduce the terrifying idea of 'super-parasites' which are drug-resistant.

....
....
....
....
....
....
....
....
....
....
....
....
....
....
....
....
....
....
....

We all know the threat caused by antibiotic-resistant bacteria. The thought of untreatable infections killing children and adults alike because the drugs no longer work strikes fear into every heart. Drug-resistant malaria parasites evolve in the same way and their potential for causing death and destruction is similar.

Is there an answer?
Scientists across the world are working day and night to find ways of controlling malaria. There has been a long-running battle between people and the parasite which causes the disease. Every time we think we have it beaten, the parasite evolves a new way of surviving.

Sometimes a simple solution is best. Instead of using drugs to kill malaria parasites, why not just keep the mosquitoes away from the people?

Unit 7
Guided

> Carry on by explaining how mosquito nets impregnated with insecticide can help prevent people getting infected.

....
....
....
....
....
....
....

In spite of nets and insecticides millions of people get malaria every year. Many people infected live in countries with malaria – but some of us come home from holiday with an uninvited guest! New drugs are hard to discover and expensive to develop and the parasites can develop immunity to a new drug in a matter of years.

> You could produce a bar chart from data in Article 3 to show how quickly the malaria parasite has become resistant to several relatively new drugs.

....
....
....
....
....
....

Scientists at the London School of Hygiene and Tropical Medicine are on the case, along with colleagues around the world. We have a new tool on our side: we can read the genome of the malaria parasites and find out exactly how they are becoming immune to our drugs.

> Use what you have discovered to write a bit more about how genome analysis can help scientists understand how the malaria parasite develops immunity to drugs and how they can use it to develop new drugs, but do not go into too much detail or your readers will get lost.

....
....
....
....
....
....

Here in the UK we protect our children against many terrible diseases by giving them vaccinations. Why don't we vaccinate people against malaria? It all comes back to the problems of the parasites – and genome analysis may help us there as well.

> You can add a bit here about how scientists are trying to develop vaccines against malaria and the problems they face.

....

....

....

....

....

What can we do?

We can't read the genome of a malaria parasite, or develop a new drug, or find a new insecticide to keep mosquitoes at bay. But we CAN support the people who do. Money is short at the moment and the politicians are cutting back on funding for research.

> Add some more information here explaining how the problems of malaria are likely to grow if we don't do something about it, and suggest some of the things we might do.

....

....

....

....

Revision task 2

This revision task has fewer hints and sample answers than Revision task 1 to give you a chance to practise answering assessment-style questions without so much support.

Revision task brief

You are provided with the following articles:

Article 1: 1.1 billion reasons why light poverty must be eradicated

Article 2: Case study of contemporary physics: ITER

Article 3: Playing catch up: can the stellarator win the race to fusion energy?

You need to gain an understanding of the scientific issue discussed in the articles so that you are able to interpret, analyse and evaluate the articles.

Reading and making notes

As for the previous task, you should make sure you read the articles carefully at least twice before you do anything else. Then before you tackle the questions, make sure you annotate the articles and make notes.

Article 1

1.1 billion reasons why light poverty must be eradicated

3:50PM BST 11 Sep 2015

By Eric Rondolat, Chief Executive Officer, Philips Lighting

Across the globe the combined populations of Birmingham and Bristol die needlessly every year through light poverty, says Eric Rondolat of Philips Lighting.

A child does homework by candlelight.

The resourcefulness of the human race never ceases to surprise. We're learning how to 3D print human organs; travelling in self-driving cars; and recently landed a spacecraft on the surface of a comet more than 500 million kilometers away.

Yet, for all of these triumphs of ingenuity, we live in a world where 1.1 billion people– more than one in seven – still do not have access to electric light.

The lack of this most fundamental service puts a stranglehold on human development. Without artificial light, life as we know it grinds to a halt at sunset. Communal life stops, children are unable to study, and businesses are forced to close.

Deprived of electric light, people resort to candles, kerosene lamps and fires to counter darkness – all too often with devastating consequences. These primitive light sources claim the lives of 1.5 million people every year through fires and respiratory illnesses – the same number killed annually by HIV related illnesses.

Light poverty and the millions of associated deaths are avoidable – the technology to balance this inequality is all around us and taken for granted across most of the world. In those countries blighted by light poverty, the difficulty lies in administering the cure, not in creating it.

At first glance, the obvious solution might appear to be for affected countries to invest in electric grids and power plants to provide a reliable energy supply – and thus light – to all their citizens. However, the geography of many developing nations makes this simply unfeasible both logistically and financially. Furthermore, in many of those communities deemed to have access to electricity, erratic grid connections regularly plunge homes and businesses into darkness without warning.

Instead, the focus should be on developing off-grid solutions that give communities – particularly in rural areas – a dependable and sustainable energy source. As recent research by the United Nations Environment Programme (UNEP) suggests, solar-powered LED lighting provides a low-cost alternative that not only alleviates light poverty but also reduces carbon emissions, indoor air pollution, and health risks.

The economic argument for solar-powered LED systems is irresistible too. For example, families in the Democratic Republic of Congo spend up to 30 per cent of their annual income on fuel for kerosene lamps. And some African households are buying lighting at the equivalent of £65 per kilowatt-hour, more than 100 times the amount paid by people in developed countries.

A single solar-powered LED lantern uses zero energy and can fill a room with electric light without any carbon emissions or noxious fumes for a one-off cost of less than £15 – compared to the £33 average annual fuel bill of running a kerosene lamp.

On a larger scale, community light centers that combine energy-efficient LED luminaires with solar panels can produce sustainable lighting for public places such as markets and sports grounds, without the need for costly infrastructure. These centers are relatively inexpensive, easy to install and maintain, and far more reliable than electric grids, which can suffer from outages in remote areas. Philips will have installed more than 100 of them in rural Africa by the end of this year.

These and other projects and initiatives, such as the Global Off-Grid Lighting Association, are helping drive down the proportion of people trapped in light poverty. The latest research by the World Bank suggests that the number of people without access to electric power fell from 1.2 billion to 1.1 billion between 2010 and 2012.

A solar lantern.

But progress is simply too slow, particularly in the worst affected areas. In Africa, the latest projections suggest that the number of people with no access to electricity is likely to rise from 600 million to 700 million by 2030.

Part of the problem lies in red tape. Inadequate and outdated regulations and subsidy schemes stifle the adoption of off-grid solutions in many parts of Africa. The introduction of VAT and tariff exemptions, together with minimum quality standards can create an environment conducive to helping cheaper off-grid technologies thrive.

Such measures, according to research by UNEP, have helped millions of low-income households beyond the grid to gain access to clean and sustainable lighting. In Ethiopia, for example, collaboration between the World Bank, the Government of Ethiopia and Lighting Africa established a US$20 million financing facility for off-grid solutions. Within 18 months, the scheme enabled more than 300,000 quality-verified solar lighting products to be imported, providing 1 million Ethiopians with access to electric light.

The benefits of intervention extend beyond those communities affected. Bringing one-seventh of the world population out of the dark would deliver a huge boost to global GDP by promoting education and business growth.

Recognizing the scale of the problem and its impact on social and economic development, the United Nations proclaimed 2015 the International Year of Light. It aims to bring together governments, private sector partners and scientific bodies, and raise awareness of the power of light-based technologies in solving global challenges in energy, education, agriculture, public infrastructure and health.

But with more than half the year gone, governments and businesses must seize this moment to embrace the technology at their disposal and immediately begin delivering lasting change to communities left in darkness.

The combined pledges from the European Commission, individual European countries and the USA during the UN Climate Summit, can halve energy poverty by 2030. But I believe we need to go further still. 60 per cent of the global population will be living in cities by the time we reach the year 2030, increasing the world population by a little over 1 billion, challenging access to secure, more sustainable, safe energy such as electricity for lighting.

Political will, combined with entrepreneurial drive, can snuff out this injustice if cemented by shared determination and vision. Without a concerted effort, the equivalent of the combined populations of Birmingham and Bristol will continue to die needlessly every year through light poverty. The time to act is now.

[Source: www.telegraph.co.uk]

Article 2

Case study of contemporary physics: ITER

Student name: Chris Short
University: ▇▇▇▇
Department: Physics   2017

ITER

ITER (International Thermonuclear Experimental Reactor), is the world's most promising facility for research into nuclear fusion power.

In southern France, 35 nations have together contributed to a building project costing $14.4 billion (as of 2015). In the ITER project they are attempting to build the world's largest tokamak (a type of fusion reactor) in the hope that they can provide carbon-free, sustainable and feasible fusion reactions that will pave the way for a new type of power station, all working on the principle that powers the Sun.

The ITER project began building in 2013 and will not be finished until at least 2020. It brings together the greatest scientific minds from the 35 countries, including China, the European Union, India, Korea, Russia, Japan and the USA, to work on a 35-year build and research facility with the aim to design a working fusion reactor at the end. The recent Brexit vote, however, has sparked a debate over the UK's inclusion in the project post-leave.

The power of a tokamak is proportional to the number of fusion reactions taking place per second in the reaction vessel. A larger reaction vessel with more contained plasma is therefore capable of producing a much higher power output.

At ten times the current largest reaction vessel (the JET reactor at CCFE) ITER's Tokamak hopes not only to simply be bigger, but also that the large vessel can contain the plasma for longer allowing more energy to be produced. ITER is aiming for: [1]

- **Production of 500 MW power output.** The current world record holders for fusion power are the Culham Centre for Fusion Energy (CCFE) with the Joint European Tokamak (JET) reactor. This reactor achieved a 16 MW output from 24 MW input, a Q-factor (measurement of efficiency) of 0.67. ITER aims for a Q-factor of at least 10 from a 50 MW input, generating a 500 MW output. The aim of the ITER tokamak is not to produce electricity. It simply has to means-test the idea that energy can be gained from this type of reactor, to pave the way for a similar machine that can actually use the energy produced to generate electricity.

- **Integrate technologies of smaller plants.** At the moment, many tokamaks are looking into very specific areas of fusion, for instance focusing specifically on the magnets used for containment or the heating methods within the vessel. ITER aims to bring the optimised technologies from different research teams together into one plant, in the hope the combined efforts make a viable, working plant.

- **Test various combinations of isotopes for fusion reactions.** Currently the main reactants in fusion are simple hydrogen–hydrogen fusion or deuterium–deuterium fusion (deuterium is a hydrogen atom with a neutron). Hydrogen and deuterium are abundant in the Earth. However, in theory at least, deuterium–tritium (a hydrogen atom with two neutrons) or even tritium–tritium fusion should be easier to self-sustain as a chain reaction. Instead of the energy being produced (relatively) slowly by a series of fusions, the energy of each reaction is released at once, maintaining the heat. The problem is that tritium is very rare, so ITER is looking into the possibility of producing tritium in the reaction vessel itself from the by-products of deuterium fusion. Part of the research focus is to try and find the best mix of 'ingredients' to use.

- **Demonstrate the safety of fusion reactors.** ITER needs to prove that it is not only feasible, but safe, to set up fusion reactions in a vessel with a view to generating electricity, as failure to successfully contain a fully-fledged fusion reaction could be catastrophic.

What is fusion?

Nuclear fusion is the process by which almost all stars produce light and radiation. The immense pressure at the centre of stars causes electrons in hydrogen atoms to separate from the nucleus forming a 'soup' of nuclei and electrons called plasma. The nuclei collide and fuse into heavier atoms like helium, and due to a property known as binding energy, a large amount of energy is released.

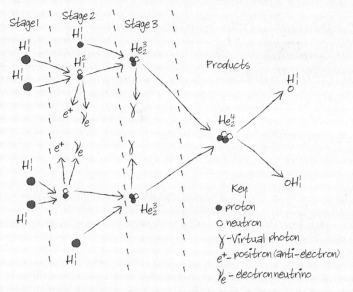

Diagram showing the reaction mechanism for a simple (no heavy isotopes) hydrogen fusion reaction.

Early research into fusion found that hydrogen fusion is the most feasible on Earth, as the lack of pressure at the surface of the Earth means that temperatures of ~150,000,000K are needed for even the simplest fusion reactions. Research also found that D–T (deuterium–tritium) reactions would be the easiest to self-sustain [2], as the energy produced helps to maintain the heat needed in the vessel.

In order to maintain fusion reactions, the vessel must be:

1 **Superheated.** This means that the vessel must be able to heat a vacuum and plasma to over 150,000,000K and the materials in the vessel must withstand this heat. Boron plating is often used, because although it is extremely expensive it is up to the job.

2 **Able to achieve high plasma density.** The plasma would not collide frequently enough if allowed to move freely in the vessel, so magnets are used to contain the charged plasma in a small volume, causing a) more frequent collisions and b) helping reduce heat damage to vessel walls. Some of the fast-moving neutrons (which are not affected by magnetic fields) can collide with the vessel walls, even getting through and damaging external parts.

3 **Able to maintain the temperature and high plasma density for extended periods of time.** The times in question are of the order of four minutes, and although the electromagnets are effective, they are not efficient – they require super-cooling (in a super-heated vessel!) and a lot of energy to be powerful enough.

The fusion vessels used are largely tokamaks (magnetic fusion), but LASER fusion is an alternative process. It is not as feasible for large scale energy production, as it uses a laser to heat the fuel and contains it in a spherical (rather than toroidal) chamber.

Tokamaks – electricity generation for the future?

Current power plants all use fossil fuels, nuclear energy, wind or hydro-electric power to develop the kinetic energy of a rotating dynamo and a coil of wire with a magnet to produce electricity, using Faraday's Law. So, for example, in combustion power generation, the burning of fossil fuels heats a tank of water. The rising steam is used to turn the turbine. Similarly, in nuclear power stations, nuclear fission produces radiation and relativistic neutrons to again heat water and turn a turbine.

Tokamaks were first developed by the Russians in the 1960s – their name comes from a Russian acronym for 'Toroidal chamber with magnetic coils'. Most scientists agree they are the most likely way in which we will be able to generate electricity from nuclear fusion. Tokamaks are still very experimental machines, but they actually use similar principles to nuclear fission generators. The energy produced in the fusion reactions is used to maintain the plasma, and escaping, fast-moving neutrons are caught by heat-shielding plates. As they get hot, they heat the water, creating steam and turning a turbine.

The tokamak uses resistive coils of wire to heat up the toroidal (doughnut-shaped) vessel until plasma is produced, then large magnets are used to contain the plasma to a volume about the same as a power cable, one at the top, one at the bottom. The plasma can be controlled because a moving charge (the positive nuclei and negative electrons) of a plasma can be influenced by magnetic fields.

Tokamaks are designed to heat plasma very efficiently, with a variety of different heating methods: [3]

- **Vacuum Vessel.** The low pressure of the vessel while under the same volume causes cooling of the vessel itself. While this may seem a hindrance, it in fact cools the magnetic coils, allowing them to operate more effectively, containing the plasma in a smaller space, so a smaller volume must be heated.
- **Plasma Currents.** This makes use of eddy currents that occur in transformers causing the cores to heat using Ohmic heating. This heat also applies to any moving charge or 'current'. Thus magnetic fields can cause heating in the plasma. This supplies about a third of the heat required.
- **Neutral Beam Injection** is the process where additional fuel is propelled at high speed into the plasma current and is ionised then slowed by the magnetic field. The sudden change in kinetic energy is released as radiation and heats the plasma
- **Radiofrequency Heating.** Specific frequencies of radio waves are chosen such that they match those where absorption in the plasma is very high. The photons are absorbed by the plasma, transferring a large amount of energy very quickly.

The biggest, most powerful tokamak current in use is the JET (Joint European Torus) which is found in the UK, at the Culham Centre for Fusion Energy.

The JET (Joint European Torus) at the Culham Centre for Fusion Technology

If and when it is completed, the ITER will be twice the size of the JET, and the plasma chamber will have about 10 times the volume of the chamber in the JET. If ITER is successful, it will bring fission powered electricity more than one step nearer – promising safer, greener and cheaper electricity for us all.

References

1. International Thermonuclear Experimental Reactor, ITER, 2017, The Way to New Energy [Online], Available from: https://www.iter.org/ [accessed June 2017]
2. Tipler, PA, 2007, Physics for Scientists and Engineers, 6th Edition, W.H. Freeman and Co.
3. Culham Centre for Fusion Energy, CCFE, 2012, Tokamaks [Online], Available from: http://www.ccfe.ac.uk/Tokamak.aspx [accessed June 2017]

Article 3

Playing catch up: can the stellarator win the race to fusion energy?

18 April 2016
by Taylor Heyman

2017

An experimental nuclear fusion technology at the Max Planck Institute of Plasma Physics produced its first hydrogen plasma in February 2016, sparking speculation that the stellarator could overtake the tokamak as the leading experimental form of nuclear fusion energy production.

The stellarator

It was the culmination of a decade's work for a dedicated group of international scientists when German Chancellor Angela Merkel stepped up to a cube-shaped crystal button at the Max Planck Institute of Plasma Physics (IPP). The Chancellor, a former physicist with a PhD in hydrocarbon research, gave a short speech and pushed the button, setting in motion the first hydrogen plasma test of the Wendelstein 7-X stellarator nuclear fusion device.

The initial test of the world's first large-scale stellarator design lasted 0.4 seconds and heated a small amount of hydrogen into plasma reaching 80,000,000°C. Crucially, a long enough time to study.

"This by far exceeds our expectations for the first operation phase of Wendelstein 7-X.", says scientific director for the project, Professor Thomas Klinger.

The test was deemed a success, and significant steps have been made from the initial trials performed in December with helium gas. "At one million watts microwave heating power we could maintain the plasma for four seconds," says Klinger. "We started in December with 0.04 seconds which means an improvement by 100x within only two months."

The future of energy production

News of the test raised hopes of a clean, safe form of nuclear energy in the not-so-distant future, and of using fusion to replace current commercial nuclear fission power stations. Nuclear fission starts with heavy nuclei that are split to release energy. The nuclei used in fission reactors are normally uranium or plutonium isotopes, as their atoms have large nuclei that are easy to split.

Unfortunately, fission reactors also create radioactive waste and can cause devastation in the event of an accident. The most recent incident was the Fukushima plant in Japan which was hit by a tsunami in 2011. The disaster saw a 20km cordon put around the site and the clean-up operation is still in progress, five years on.

The process is the same as the way the sun and stars create energy

Nuclear fusion, on the other hand, works quite differently. The process is the same as the way the sun and stars create energy; the challenge is how to recreate these conditions on earth. Instead of making the nuclei lighter, as with fission, it works by fusing light atoms into heavier ones thereby converting mass into energy. The most common isotopes used in this process are deuterium and tritium, which are fused with each other to create helium and a neutron. As the two atoms join, they lose a small amount of their mass in the form of energy, at a level greater than that required to create the fusion.

This is the embodiment of Einstein's famous $E=mc^2$ formula; the lost mass, multiplied by the square of the speed of light equals a large amount of energy.

It sounds so simple; however, recreating this atomic process on earth has been the life's work of many dedicated scientists. The sun is able to create so much energy because its gravitational force is so large. To replicate this on earth, scientists need to create an environment of 150,000,000°C to force the fusion of atoms. Reaching the temperature is not a problem, according to Professor Tony Donné, programme manager at EUROfusion, a body which co-ordinates and funds European fusion research activities, and is made up of research organisations and universities from 26 European states and Switzerland. "We are now trying to study how can we make hydrogen hot enough that we get fusion, how do we keep it stable, how do we extract the electricity from it," he says.

The main problem to solve with nuclear fusion is how to keep the gas, which turns to plasma when heated to a high temperature, from touching the sides of the reactor. As the atoms fall apart into positive nuclei and negative electrons during the process, the plasma can be controlled by magnetic fields to keep it floating around the inside of the reactor. It is this magnetic process that is different in the two main reactor designs in this field, the tokamak and the stellarator.

To put it simply, the tokamak confines the plasma by running a magnetic current through it to induce a magnetic field, whereas the stellarator has a complex group of external magnetic coils to control the plasma from the outside of the reactor.

Tokamak or stellarator?

Until recently, the tokomak design, invented in the 1960s, has been prevalent in experiments and studies in the field. However, the magnet design does cause issues preventing the system from running continuously without auxiliary assistance. "In a tokamak, plasma is prone to some instabilities, and we have some ways of dealing with this, but it would be better if we could have the magnetic field entirely external," explains Donné, whose organisation co-ordinates nuclear fusion projects from JET, the world's biggest tokamak design fusion reactor based in the UK, to ITER, a collaboration between 35 different countries to build an even larger tokamak machine based in France, to Wendelstein itself.

This is where developments in the stellarator come in. The design was first developed in 1950 by Lyman Spitzer at what would later become the Princeton Plasma Physics Laboratory. It was popular in the next two decades, but as the tokamak design began to show better results, the stellarator faded. Recent developments in technology have led to the return of the design. "It is only now we have very powerful computers that we can design this type of device" says Donné.

It is not as simple as replacing the tokamak design with the stellarator for future investigation, for one thing, the stellarator is generations behind the tokamak in terms of progressing toward energy production.

Comparing Wendelstein to the largest tokamak project in operation today, Donné thinks there is a way to go. "At JET we have all the nuclear licences. Wendelstein doesn't have a nuclear licence, so they are only allowed to work with hydrogen, helium or deuterium," he says. The team currently working on Wendelstein do not list proving energy production as one of their aims. This will come later for another group of scientists, or as a follow up to the Wenselstein 7-X study.

This doesn't mean the stellarator has nothing to contribute. Its predicted ability to perform very long pulses will enable the team to study the plasma's behaviour in the reactor over a period of thirty minutes, compared with just one minute in the tokamak. This knowledge will be shared with tokamak scientists around the world to freely incorporate into their own work.

Klinger remains positive about Wendelstein's potential to contend with the tokamak. "Thanks to magnetic field optimisation Wendelstein 7-X has a good chance to catch up with the best performing tokamaks we have," he says. "We have made the first steps, but already those look very promising."

When will the lights be turned on by fusion?

The first hydrogen test was an auspicious day for the 500-strong WendelStein 7-X team, but there is still considerable work to do. The plasma vessel has just been reopened to install carbon tiles to protect the walls, with the aim of extending the plasma duration to 30 minutes, a first for a fusion reactor.

The team is working with industry partners rather than manufacturing the parts for the stellarator itself. This will ensure that when the device is ready to be used on a commercial scale, industry organisations will be ready to help.

According to Donné, fusion energy production could be in use as soon as the second half of this century. EUROfusion is currently in the pre-conceptual design phase of DEMO, a demonstration fusion reactor that will produce energy on a large scale and convert it to electricity. Presently, the idea is to use the tokamak design, but this could change, depending on a variety of factors.

ITER, the tokamak reactor being built in France, will come online around 2025 says Donné. It will begin to use the tritium and deuterium to create energy around 2035. The facility will be experimental, and lessons learned from it will be taken onward to build DEMO.

The DEMO team is taking an open view to the technologies it will use, although tokamak is currently the preferred design, with the Wendelstein stellarator serving as a back-up. "If we run into a problem with the Tokamak and we cannot make it work, then the stellarator can be a way out," explains Donné. "It could also be that at one point we decide that DEMO will not be a device on the tokamak line but the stellarator line."

"With better funding the worldwide fusion community could make more progress."

Both Klinger and Donné are agreed that progress would be quicker with better funding. "With better funding the worldwide fusion community could make more progress just by working more in parallel than in series, like it is now the case," says Klinger. "There are too many old fusion devices and too few new machines or machines under construction."

It's a long and bumpy road ahead for nuclear fusion scientists working to make fusion energy a reality, but recent developments and funding commitments mean they have a map, at least. X marks the spot - clean energy for all.

[Source: www.power-technology.com]

Unit 7

Answering questions

Answer all the questions in the spaces provided.

1 Discuss the implications of the scientific issue identified in the articles.

> You should already have made notes based on the articles. Before you start writing your answer, look back at your notes to remind yourself about the main issue in the three articles. Make sure you refer to the articles and to other reading you may have done. Make a plan of what you want to say to ensure your answer is clear, coherent and logical.
>
> Make sure you refer to all of the different articles in answering this question. You should draw a wide range of links to and between the ethical, social, economic and/or environmental implications of the science.

Unit 7

Unit 7

2 Identify the different organisations/individuals mentioned in the articles and suggest how they may have had an influence on the main scientific issues.

> You should look back at your notes to remind yourself about the **organisations** and **individuals** mentioned in the articles.
>
> If you have made good notes, you will have these organisations and individuals already identified which will make answering this type of question much easier, perhaps in a table or bullet points.
>
> In your answer you need to make clear links between the institutions and people mentioned and the original articles.

3 Discuss whether Article 2 has made valid judgements.

In your answer, you should consider:
- how the article has interpreted and analysed the scientific information to support the conclusions/judgements being made
- the validity and reliability of data
- references to other sources of information.

> Your notes should help you to plan a clear and well-structured answer to this question. Make sure you take the time to plan a clear, coherent and logical answer.
>
> Make sure you refer to the points made in the bullet points given to guide your answer. You might like to structure your answer in three sections to make sure you do this.
>
> Look at your notes to help you highlight the main points you need to answer this question well.

Unit 7

4 Suggest potential areas for further development and/or research of the scientific issue from the three articles.

> In your notes you may have included thoughts on potential future development and research. In this answer you can suggest more than one direction for future R&D. You will need to justify your choices by explaining why that research is needed.
>
> You are expected to link your ideas for further development and research to science from the three articles. Make sure you include reference to all the articles, not just one or two.

5 Following the 2015 UN Year of Light, it has been decided to have a UK Year of Light.

You have been asked to produce a report to go into every school in the country, to be used in assemblies and science lessons with students aged 13–16.

The aim of the article is to make students aware of some of the issues surrounding the provision of light around the world. These sessions will be linked to fund-raising efforts to try and make sure that by 2050 everyone has access to the lighting they need, produced in ways which will not harm the environment.

Use information from the three articles provided for this task.

When writing your article, you should consider:
- who is likely to read the article
- what you would like the reader to learn from the article.

> You should plan your answer and make sure you consider the following points:
> - Keep your target audience in mind.
> - Plan what you want them to know.
> - Decide how you can include good, valid, reliable science but still make your article interesting – you want people to keep reading to the end.
> - A good headline will catch your reader's attention.
> - Use information from all the articles to show the examiners that you have read and taken in information from all your sources and understand where they agree and where they take different approaches.
> - If you can, include some extra information from your reading around.
> - Keep your tone, style and level of scientific terminology the same throughout the article.

The periodic table of elements

Key

| relative atomic mass |
| atomic symbol |
| name |
| atomic (proton) number |

Example: 1.0 H hydrogen 1

1	2											3	4	5	6	7	0 (8)
(1)	(2)											(13)	(14)	(15)	(16)	(17)	(18)
																	4.0 He helium 2
6.9 Li lithium 3	9.0 Be beryllium 4											10.8 B boron 5	12.0 C carbon 6	14.0 N nitrogen 7	16.0 O oxygen 8	19.0 F fluorine 9	20.2 Ne neon 10
23.0 Na sodium 11	24.3 Mg magnesium 12	(3)	(4)	(5)	(6)	(7)	(8)	(9)	(10)	(11)	(12)	27.0 Al aluminium 13	28.1 Si silicon 14	31.0 P phosphorus 15	32.1 S sulfur 16	35.5 Cl chlorine 17	39.9 Ar argon 18
39.1 K potassium 19	40.1 Ca calcium 20	45.0 Sc scandium 21	47.9 Ti titanium 22	50.9 V vanadium 23	52.0 Cr chromium 24	54.9 Mn manganese 25	55.8 Fe iron 26	58.9 Co cobalt 27	58.7 Ni nickel 28	63.5 Cu copper 29	65.4 Zn zinc 30	69.7 Ga gallium 31	72.6 Ge germanium 32	74.9 As arsenic 33	79.0 Se selenium 34	79.9 Br bromine 35	83.8 Kr krypton 36
85.5 Rb rubidium 37	87.6 Sr strontium 38	88.9 Y yttrium 39	91.2 Zr zirconium 40	92.9 Nb niobium 41	95.9 Mo molybdenum 42	(98) Tc technetium 43	101.2 Ru ruthenium 44	102.9 Rh rhodium 45	106.4 Pd palladium 46	107.9 Ag silver 47	112.4 Cd cadmium 48	114.8 In indium 49	118.7 Sn tin 50	121.8 Sb antimony 51	127.6 Te tellurium 52	126.9 I iodine 53	131.3 Xe xenon 54
132.9 Cs caesium 55	137.3 Ba barium 56	138.9 La* lanthanium 57	178.5 Hf hafnium 72	180.9 Ta tantalum 73	183.8 W tungsten 74	186.2 Re rhenium 75	190.2 Os osmium 76	192.2 Ir iridium 77	195.1 Pt platinum 78	197.0 Au gold 79	200.6 Hg mercury 80	204.4 Tl thallium 81	207.2 Pb lead 82	209.0 Bi bismuth 83	(209) Po polonium 84	(210) At astatine 85	(222) Rn radon 86
(233) Fr francium 87	(226) Ra radium 88	(227) Ac* actinium 89	(261) Rf rutherfordium 104	(262) Db dubnium 105	(266) Sg seaborgium 106	(264) Bh bohrium 107	(277) Hs hassium 108	(268) Mt meitnerium 109	(271) Ds darmstadtium 110	(272) Rg roentgenium 111							

Elements with atomic numbers 112–116 have been reported but not fully authenticated

*Lanthanide series

| 140 Ce cerium 58 | 141 Pr praseodymium 59 | 144 Nd neodymium 60 | (147) Pm promethium 61 | 150 Sm samarium 62 | 152 Eu europium 63 | 157 Gd gadolinium 64 | 159 Tb terbium 65 | 163 Dy dysprosium 66 | 165 Ho holium 67 | 167 Er erbium 68 | 169 Tm thulium 69 | 173 Yb ytterbium 70 | 175 Lu lutetium 71 |

*Actinide series

| 232 Th thorium 90 | (231) Pa protactinium 91 | 238 U uranium 92 | (237) Np neptunium 93 | (242) Pu plutonium 94 | (243) Am americium 95 | (247) Cm curium 96 | (245) Bk berkelium 97 | (251) Cf californium 98 | (254) Es einsteinium 99 | (253) Fm fermium 100 | (256) Md mendeleevium 101 | (254) No nobelium 102 | (257) Lr lawrencium 103 |

Unit 1 formula sheet

Wave speed $\qquad v = f\lambda$

Speed of a transverse wave on a string $\qquad v = \sqrt{\dfrac{T}{\mu}}$

Refractive index $\qquad n = \dfrac{c}{v} = \dfrac{\sin i}{\sin r}$

Critical angle $\qquad \sin C = \dfrac{1}{n}$

Inverse square law in relation to the intensity of a wave $\qquad I = \dfrac{k}{r^2}$

Unit 5 formula sheet

Work	$\Delta W = F\Delta s$
Work done by a gas	$\Delta W = p\Delta v$
Efficiency	$\text{efficiency} = \dfrac{\text{useful energy output}}{\text{total energy input}}$
Efficiency for heat engines	$\text{efficiency} = 1 - \dfrac{Q_{out}}{Q_{in}}$
Maximum theoretical efficiency	$\text{efficiency} = 1 - \dfrac{T_C}{T_H}$

Thermodynamics

Ideal gas equation	$pV = NkT$
First Law of Thermodynamics	$Q = \Delta U + W$
Specific heat capacity	$Q = mc\Delta T$
Specific latent heat	$Q = \Delta mL$

Materials

Density	$\rho = \dfrac{m}{v}$
Hooke's law	$F = k\Delta x$
Young's modulus	Stress $\sigma = \dfrac{F}{A}$
	Strain $\varepsilon = \dfrac{\Delta x}{x}$
	$E = \dfrac{\sigma}{\varepsilon}$
Elastic strain energy	$\Delta E_{el} = \dfrac{1}{2}F\Delta x$

Answers

Unit 1: Principles and Applications of Science I

Revision test 1

Section A: Structures and functions of cells and tissues
(pages 2–7)

1. (a) A (1)
   (b) 0.25 ms (allow up to 0.5 ms) (1)
   (c) 2 ms (1)
   (d) A (1)
   (e) The Na^+/K^+ ATP pumps use energy to rapidly transport Na^+ ions out of the cell (1) and K^+ into the cell. (1)
2. (a) eyepiece lens = total magnification/objective lens (1) = 500/10 (1) = × 50 (1)
   (b) size of real object = size of image/magnification (1) 12 000/500 (1) = 24 µm (1)
   (c) Aquatic plants need to ensure their leaves are as close to the surface of the water as possible (1) so they can get as much light as possible for photosynthesis (1). The air spaces make the plant stems more buoyant so they can hold the leaf surface on the water surface. (1)
3. (a) A specialised cell has structural adaptations (1) to help it carry out a particular function (1). [or words to that effect]
   (b) (i) Cell A is a palisade cell, it carries out photosynthesis. (1) Cell B is a root hair cell, it takes up water and mineral salts. (1)
   (ii) Must have:
   - Cell A and Cell B both have mostly the same cell features (cell wall, plasma membrane, nucleus, vacuole, cytoplasm, mitochondria, Golgi apparatus, endoplasmic reticulum (smooth and rough), ribosomes). (1)

   Three of the following: (3)
   - They both have large vacuoles, cell A to help keep the cell rigid and cell B to help with the flow of water into the cell.
   - Cell A has many chloroplasts, so the cell can carry out photosynthesis.
   - Cell B does not have any because its main function is to take up water and mineral salts and the roots are under the ground and are not exposed to light, so having chloroplasts would be a waste of cell resources.
   - Cell A is a uniform cylindrical shape to ensure the chloroplasts are in the best position to absorb light.
   - Cell B has thin protrusions, which grow between soil particles so the root cell is in contact with water molecules.
   - Cell B has a large surface area to volume ratio.
   - Cell B also has thin cell walls so that water can pass through the cell wall easily by osmosis.
4. (a) B
   (b) Within the muscle cells are myofibrils, which are bundles of protein filaments (1). Each myofibril is made up of repeating units called sarcomeres. (1)
5. Example answer: (6):
   - Normal red blood cells are elastic.
   - They can squeeze through capillaries.
   - Sickle cells are inelastic.
   - So they might get stuck or clump.
   - Sickle cells do not last very long.
   - So there will be a lack of red blood cells in the body.
   - Red blood cells carry oxygen to cells for respiration, which releases energy.
   - A lack of red blood cells would mean less energy and the person would feel tired and may be short of breath.

Section B: Periodicity and properties of elements
(pages 8–12)

1. (a) B, ^{79}Br has 44 neutrons (1)
   (b) B and E. They contain the same number of electrons and the same number of protons. (2)
   (c) (i) $(1s^2\ 2s^2\ 2p^6\ 3s^2\ 3p^6)\ 3d^{10}\ 4s^2\ 4p^5$ (1)
   (ii) The outermost electrons are in a p sub-shell. (1)
   (d) The bromine atom has a partial positive charge ($\delta+$) and the fluorine atom has a partial negative charge ($\delta-$). (1)
   (e) (i) The formula that gives the exact number of atoms (1) of each element in one molecule (1)
   (ii) $P_4 + 6Br_2 \rightarrow 4PBr_3$ (1)
   (iii) P (16.2 ÷ 31.0) = 0.523 **and** Br (83.8 ÷ 79.9) = 1.05 (1) 0.523 : 1.05 = 1 : 2, therefore the empirical formula is PBr_2 (1)
   (iv) Identity = P_2Br_4 (1)
2. (a) The energy required to remove one electron from each atom (1) in a mole of atoms (1) in the gaseous state (1).
   (b) The nuclear charge increases from Na to Ar (1). However, the shielding of the outer electrons does not change significantly as the extra electrons are added to the same quantum shell (1). Hence the outer electrons are attracted more strongly to the nucleus (1).
   (c) The outer electron removed from the aluminium atom is in a p-orbital (1) and it therefore has a higher energy than the outer s-orbital electron of the magnesium atom (1).
   (d) The outer electron of the potassium atom is at a higher energy level because it is in a different quantum shell that is further from the nucleus (1). The outer electron in the potassium atom also experiences more shielding (1) because there are more inner quantum shells in the potassium atom (than in the argon atom) (1). Both of these factors offset the higher nuclear charge of the potassium atom (1).
3. Example answer (6):
   - Electron density fluctuates in a molecule.
   - This creates an instantaneous dipole in the molecule.
   - The instantaneous dipole induces a dipole in a neighbouring molecule.
   - The two molecules attract one another.
   - The magnitude of the attraction between the molecules is dependent on the number of electrons in the molecules.
   - The greater the number of electrons, the larger the instantaneous and induced dipoles.
   - The chlorine molecule has the lowest number of electrons and hence the weakest forces of attraction.
   - The iodine molecule has the greatest number of electrons and hence the strongest forces of attraction.
   - The stronger the forces of attraction between the molecules the greater the energy required to overcome the forces
   - The stronger the forces of attraction between the molecules the higher the boiling temperature.

Section C: Waves in communication
(pages 13–19)

1. (a) B, 2.5 GHz (1)
   (b) C, microwave (1)
   (c) Each mobile phone operator is allocated a separate set of frequency bands by the **telecommunications licensing authorities**. (1) The upload signal from a mobile handset is broadcast on a slightly different frequency from the **download signal.** (1) The cell masts in adjacent cells **always use different frequencies.** (1)
   (d) $(3\ km/1\ km)^2$ = 9 times stronger (2)

2. (a) Longitudinal wave (1)
   (b) Drawing shows node at closed end (1), antinode at the open end (1)
   (c) So ¾ λ = 0.75. Therefore λ = 1.00 m (2)
3. (a) (i) Wave equation $v = f\lambda$, if f doubles with λ fixed then $v_2/v_1 = 2$ (1). Substituting into wave speed formula $v = (T/\mu)^{1/2}$ to give $T_2/T_1 = (v_2/v_1)^2$. So tension T needs to be 4 times larger (1)
   (ii) String: transverse, stationary wave (1); sound: progressive longitudinal wave (1)
   (iii) They have the same frequency (1)
   (b) Both depend on the specific frequencies at which stationary wave resonance can occur: electrons have fixed energy levels that depend on the binding forces in the atom; and the violin similarly has certain fixed **frequencies** that depend on the length, **tension** and **weight (mass / unit length)** of the string. (1)
   Violin sound notes have the same **frequency** as the standing wave in the string, but the light emitted from an atom is not the same frequency as an electron standing wave mode. Its frequency depends on the energy difference between two separate electron wave modes, i.e. the gap between two electron energy levels. (1)
4. (a) $1/n = 0.676$ (1); $C = \sin^{-1} 0.676 = 42.5°$ (1)
   (b) Path drawn shows more than one reflection and each has angle i = angle r (1) and > 42.5° (1), for example:

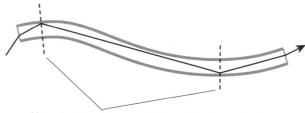

normal lines help to show angle i = angle r for each reflection but are not essential to gain marks.

   (c) Most of the light that enters the fibre will strike the outside surface at an angle **greater** (1) than the critical angle and so will be totally **internally reflected** (1). After that the ray keeps being reflected every time it hits the outside surface until finally it exits the cut end of the fibre.
   (d) Example answer (6):
   The answer should include some of the following terms:
   Similarities:
   - Signals contained inside optical fibres by total internal reflection.
   Differences:
   - Frequencies: endoscope – visible light; broadband – infrared.
   - Signal type: endoscope – analogue image, one pixel per fibre; broadband – digitised data encoded as binary (on–off flashes).
   - Data capacity: broadband: multiplexing – i.e. separate data channels down the same fibre using different frequencies; endoscopy: full colour range used for just one signal.
   - Signal range and fibre type: endoscopy – just a few metres, so ordinary (multimode) fibres are OK; broadband – hundreds of kilometres, so must use narrower 'monomode' fibres to avoid signal degradation.
   - Timing: endoscopy: immediate, synchronous image; broadband: asynchronous data with a time delay, but can be processed or stored on computers.

Revision test 2

Section A: Structures and functions of cells and tissues

(pages 20–25)

1. (a) Three from the following (3):
       plant cells have, but animal cells do not: cell wall, chloroplast, vacuole, tonoplast, amyloplast, middle lamella; animal cells have centrioles but plant cells do not.
   (b) size of real object = size of image/ magnification (1)
       = 8 mm/20 000 (1)
       = 0.0004 mm = 4 μm (1)
2. (a) A, alveoli (1)
   (b) Squamous epithelium cells cells are very flat (1), this means there is a small diffusion distance, which makes it easy for gas exchange to occur into and out of the lung. (1)
       They produce surfactant (1), this is important because it decreases the surface tension, making it easier for the alveoli to remain inflated. (1)
   (c) People with emphysema have lost some alveolar cells (1), so there are fewer cells for oxygen and carbon dioxide to diffuse through and a decrease in surfactant (1), so the alveoli are more likely to collapse and stick together, further reducing the surface area for diffusion. (1)
3. (a)

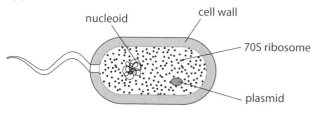

(3)

   (b) Gram-negative bacteria when stained with crystal violet and then washed with acetone and absolute alcohol (1) do not retain the stain, whilst Gram-positive bacteria do. (1)
   (c) This is because Gram-negative bacteria have an outer layer. (1)
4. (a) (i) B, node of Ranvier (1)
       (ii) D, nucleus of Schwann cell (1)
   (b) The myelin sheath is an insulating layer (1) that enables the electrical impulse to travel faster along the nerve. (1)
   (c) Example answer: (6)
       - Synapses are gaps between neurons, which action potentials have to cross.
       - The electrical impulse/signal/action potential arrives at the end of the pre-synaptic membrane.
       - The pre-synaptic membrane depolarises.
       - Calcium channels open and calcium ions flow into the neuron.
       - Calcium ions allow the vesicles containing the neuro-transmitter to fuse with the pre-synaptic membrane.
       - Neurotransmitter is released into the synaptic cleft/ gap.
       - Neurotransmitter binds with receptors on the post-synaptic membrane.
       - Sodium channels open and sodium ions flow through the channels.
       - The membrane depolarises and starts an action potential.
       - The post-synaptic membrane receptors will release the neurotransmitter and it can be taken up again by the pre-synaptic membrane and be reused or it can diffuse away and be broken down.

Section B: Periodicity and properties of elements
(pages 26–30)

1. (a) (i) C (1)
      (ii) A (1)
   (b) The atom of a group 2 element loses two electrons (1) from its outer s-orbital (1)
2. (a) B (1)
   (b) (i)

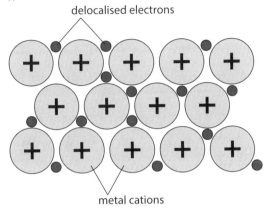

Diagram showing at least three layers of cations and electrons (1)
Correct labels (1)

   (ii) The electrostatic force of attraction (1) between metal cations and the delocalised electrons. (1)
   (c) (i) Magnesium has two delocalised electrons per ion and sodium has only one delocalised electron per ion / magnesium has more delocalised electrons per ion. (1)
   The magnesium ion has a greater charge than the sodium ion. (1)
   The force of attraction (1) between the delocalised electrons and the cations is greater in magnesium than in sodium. (1)

3. (a) Ba changes from 0 to +2. This is oxidation. H changes from +1 to 0. This is reduction. (1)
   The reaction is a redox reaction because both reduction and oxidation are taking place. (1)
   (b) (i) $(2.74 \div 137) = 0.0200$ (mol) (1)
      (ii) $n(H_2) = 0.0200$ mol (1)
         vol.$(H_2) = 0.0200 \times 24\,000\,cm^3 = 480\,cm^3$ (1)
      (iii) $n(Ba(OH)_2) = 0.0200$ mol (1)
         $[Ba(OH)_2] = 0.0200 \times (1\,000 \div 250)$
         $= 0.0800\,mol\,dm^{-3}$ (1)

4. (a) The chlorine compounds used are toxic (1) and hence they kill any bacteria in the water (1).
   (b) Any two from:
      - as flame retardants
      - added to furniture foam
      - as plastic casings for electronics
      - in halon fire extinguishers
      - in film photography.
   (c) Example answer (6);
      - Add aqueous chlorine to aqueous sodium bromide.
      - Colourless solution of sodium bromide turns orange.
      - Bromine is displaced therefore chlorine is more reactive than bromine.
      - Add aqueous chlorine to aqueous sodium iodide.
      - Colourless solution of sodium iodide turns brown.
      - Iodine is displaced therefore chlorine is more reactive than iodine.
      - Add aqueous bromine to aqueous sodium iodide,
      - Colourless solution of sodium iodide turns brown.
      - Iodine is displaced therefore bromine is more reactive than iodine.

Section C: Waves in communication
(pages 31–38)

1. (a) (i) Two from: Higher sampling sensitivity requires more bits in each data item (1); higher sampling rate produces more data items (1); both improve quality of sound reproduction (1).
      (ii) Three of: can carry more data; can maintain quality over long distances; less affected by noise; can be stored and processed by computers.
   (b) Bit rate must be lower (typically >100 times) than the broadcast carrier wave frequency (1).
   So higher frequencies give better data speeds (1).
   Infrared is strongly absorbed by the atmosphere (1).

2. (a) Time measurements taken from graph: period = 40 ms, between two waves = 6.5 ms (1) 6.5/40 = 0.16 of a cycle (1)
   (b) (i) Sketch: with positive peaks at ~14 ms and ~54 ms and a negative peak at ~34 ms (1)

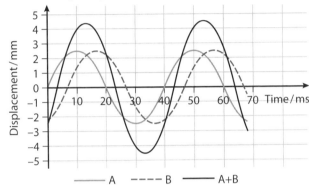

      (ii) Amplitude = 4.4 mm (values between 4.2 and 4.6 mm are acceptable, but you must give the units) (1)
   (c) (i) Antinodes occur where the two waves are exactly in phase (1); because there is a phase difference this is not the position of an antinode (1).
      (ii) Resultant amplitude at an antinode would be equal to the sum of the two component wave amplitudes, i.e. 5.0 mm (1).

3. (a) Rearrangement: $\mu = T/v^2$ (1);
   $\mu_2/\mu_1 = (v_1/v_2)^2 = (f_1/f_2)^2$; $(f_1/f_2) = 4$;
   So, $\mu_2 = 4^2 \times 1.0\,g\,m^{-1} = 16\,g\,m^{-1}$ (1)
   (b) (i) Sketch: showing nodes at both ends of the string and at the centre point – 1 whole wavelength fitting into the string (1)

      (ii) Explanation: the finger damps vibration at the centre of the string, and so prevents it establishing its fundamental mode, which would have an antinode there (1); so all the vibrational energy goes into the second harmonic mode. (1)
   (c) (i) 6.07 ms (1)
      (ii) 2.09 m (1)

4. (a) Use constancy of speed of light in wave equation $c = f\lambda$, (either by calculating $c = 508\,nm \times 590\,THz$
   $= 2.997 \times 10^8\,m\,s^{-1}$ or from memory $c = 3.00 \times 10^8\,m\,s^{-1}$) (1)
   Use that formula to obtain: f(He-Ne) = 473 THz and λ(solid state) = 532 nm (1)
   (b) Wave sources having the same frequency (1) and a fixed phase relationship (1).

(c) Example answer (4):
Wavelength and frequency:
- Linewidth for the most coherent laser (low-power continuous wave type) is very sharp indeed – less than 2×10^{-12} of the wave frequency.
- Linewidth for the least coherent laser (high-gain high-power pulsed type) is quite wide – 1% to 2% of the wave frequency.

Identifying factors affecting linewidth:
- The pulsed lasers have much higher linewidths than the continuous wave lasers.
- The higher power lasers within each laser type have higher linewidth.

Discussion of 'coherent' and 'incoherent' light sources:
- Linewidth shows coherence to be a relative measure like large and small not 'on/off'.
- Lasers usually classed as 'coherent' sources but still show a considerable range of linewidths.
- Some other sources have so little coherence – such large linewidth – that interference or diffraction patterns are no longer visible. Calling these 'incoherent' makes some sense.

Explaining which light sources will produce sharp diffraction grating spectra:
- They require monochromatic light (single frequency) – temporal coherence.
- They require a narrow beam (limited path difference) – spatial coherence.

Unit 3: Science Investigation Skills

Revision task 1

Section 1

(pages 41–46)

1 (a) Results table containing
- Suitable headings with units and measurements consistently recorded to the same precision (1)
- Indication of repeats (1)
- Indication of averages/removing anomalous results (1)

pH	Time taken for iodine to turn orange (seconds)			
	1	2	3	Mean
3	470	490	450	470
4	260	230	230	240
7	60	80	70	70
10	420	450	450	440

(b) When the iodine was first added the mixture was **blue-black** (or **black-blue**). (1)

(c) pH 7
60 + 80 + 70 = 210
210 ÷ 3 = 70
1 ÷ 70 = 0.014
(Show your working for the rest of the answer as in the worked example.)
pH 3 – 0.002
pH 4 – 0.004
pH 10 – 0.002
4 correct, (3); 3 correct (2); 2 correct (1)
Answers to 3 dp

(d) Mean = (470 + 490 + 450) / 3 = 470 (1)
For each repeat, subtract the mean and square the result
$470 - 470 = 0^2 = 0$, $490 - 470 = 20^2 = 400$, $450 - 470 = -20^2 = 400$ (1)
Add these values together = 400 + 0 + 400 = 800 (1)
Divide by the sample number – 1 = 800 / (3 – 1)
= 800 / 2 = 400 (1)
Square root this answer to give standard deviation
= √400 = 20 (1)
Divide by square root of number of repeats to give the standard error = 20 / √3 = 11.6 (anything in the range 11.5–11.7 would be acceptable) (1)

(e)
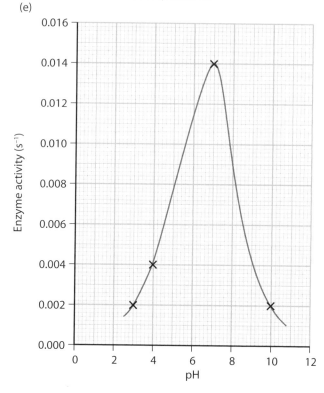

Marks given for: labels and units for both axes (1); all points correctly plotted and suitable curve (1); suitable scales (1)

(f) Amylase breaks down starch between pH 3 and 10 (1); Rate of enzyme activity increases up to pH 7 and then **decreases** (1); **Optimum pH for amylase is 7** (1).

(g) (i) 0.5 × 100/4 = 12.5% (1)
(ii) 0.07% (1)
(iii) 0.05 × 100/70 = 0.07% (1)

(h) The percentage error is largest for the measuring cylinder (1). This means that each of the volumes can be between 3.5 and 4.5 (1).

2 (a) pH 5 – any number between 100 and 200; pH 8 – any number between 100 and 250 (1). Both required for 1 mark.

(b) The optimum pH is around pH 6. This is when the time taken is lowest (1).

(c) They may have used larger volumes of starch or **less amylase** (1). The amylase may not have been so fresh so **it takes longer to work** (1). The equipment may be **less accurate** (1).

(d) The table shows that
- amlyase takes longer to break down starch but it is not known how long it takes (1)
- secondary data show that amylase does work at pH 12, just very slowly (1)
- There are no results for pH above 12 so cannot say whether unit 5 it will work at a pH above 12 (1).

3 (a) (i) Any two from
- Enzymes are affected by temperature.
- Changing the temperature can change the time taken.
- Optimum temperature for amylase is 37 °C.

(ii) The volume of amylase had to be controlled which in turn controlled the concentration of **enzyme** (1). This was done by **using a measuring cylinder to measure 4 cm³** (1).

(b) Repeat at different temperature for all pH (1), to test if amylase behaves the same way at different temperatures (1). Carry out test across full range of pH, 1–14 (1), to check all fit the pattern (1).

It is also acceptable to say investigate smaller increments of pH between 1 and 14 (1).

Section 2
(pages 47–51)

4 Your plan should include or refer to:
- a hypothesis
- equipment, techniques and /or procedures
- risks
- control variables
- dependent variables – how they will be measured, units and the precision of measurements to be taken
- independent variable – the range of measurements/ categories to be used and how they will be measured, the intervals to take measurements
- data analysis.

Example answer (12):

Hypothesis: it is predicted that the bigger the agar jelly cube the longer it takes for the dye to diffuse into the middle. This is because bigger cubes have a smaller surface area to volume ratio and a bigger distance for the dye to diffuse across so it would take longer to reach the middle.

Equipment:
- five different sized agar jelly cubes (side length: 0.5 cm, 1.0 cm, 2.0 cm, 3.0 cm, 5.0 cm) (three cubes of each size will be needed to carry out repeats)
- dye (use the same concentration for each cube and same volume)
- 250 cm³ beakers to put the cubes and dye in
- stopclock to time how long it takes for the dye to reach the middle of each cube
- ruler to measure the size of the cubes.

Method:
- Use a measuring cylinder to add 200 cm³ of the dye solution to a beaker.
- Select the first cube and confirm the length of one of the sides using a ruler.
- Place the cube into the dye and immediately start the stopclock.
- Record how long it takes for the dye to reach the middle of the cube. Record the time to the nearest 0.01 second.
- Repeat using a new cube, beaker and dye. Repeat for each size of cube three times and calculate a mean time for each size cube.
- Calculate the rate of diffusion by 1 / time.
- For each size cube calculate the surface area to volume ratio.
- Repeat for the different sized cubes.
- Draw a graph of surface area to volume ratio against rate of diffusion.

Controlled variables:
- Control the concentration of the dye using the same concentration each time.
- Control the volume of the dye by measuring 200 cm³ each time.
- Control the temperature by carrying out the whole experiment in the same room on the same day.

Health and safety
- Goggles and laboratory coat should be worn.
- Gloves could be worn if the dye being used will dye skin. Refer to instructions on the dye packaging.
- Take care with glass, cleaning up any breakages straight away and disposing of broken glass in the designated bin.

5 Your answer should include some or all of the points below (8):
- It is not clear how temperature is investigated as the method only states 'room temperature'.
- It is not clear if temperature is controlled during the procedure.
- Readings are not repeated.
- Concentrations of iodine and starch are not given.
- It is not clear if the concentration or volumes of iodine and starch are controlled and if so how they are controlled.
- Measurements are not recorded to the same degree of precision.
- It is not clear when timing will start.
- It is not clear what the end point is.
- Non-uniform temperature range (this is potentially not a problem if a graph is drawn, but a uniform range is more usual).
- Data partially support conclusion.
- Insufficient data at 70 °C and above to fully support the conclusion.

Revision task 2
Section 1
(pages 53–59)

1 (a) Example answer (3):

Voltage (V)	Current (amps)			
	1	2	3	Mean
2	1.8	1.9	2.1	1.9
4	3.7	3.7	3.8	3.7
6	5.9	6.0	9.3	6.0
8	7.6	7.7	7.8	7.7
10	8.4	8.6	8.8	8.6
12	9.4	9.2	9.3	9.3

(b) The power supply was checked for damage to the cord or plug before being used (1)
or
I was careful not to touch the bulb during the experiment as it would get hot (1).

(c) power = voltage × current or $P = VI$ (1)
All power calculated correctly using average current to 1 or 2 dp (1).
2V – 3.87 W
4V – 14.80 W
6V – 36.00 W
8V – 61.60 W
10V – 86.00 W
12V – 111.60 W

(d) 2V = 0.95 Ω = 9.5 × 10⁻¹ Ω
4V = 0.93 Ω = 9.3 × 10⁻¹ Ω
6V = 1.00 Ω
8V = 0.98 Ω = 9.8 × 10⁻¹ Ω
10V = 0.86 Ω = 8.6 × 10⁻¹ Ω
12V = 0.78 Ω = 7.8 × 10⁻¹ Ω
Transformation of equation to resistance = current / voltage (1); correct substitution and answer for all (max 2 marks) (if 4 or 5 answers are correct 1 mark)
All answers correctly in standard form (max 2 marks) (if 4 or 5 are correct 1 mark).

(e)

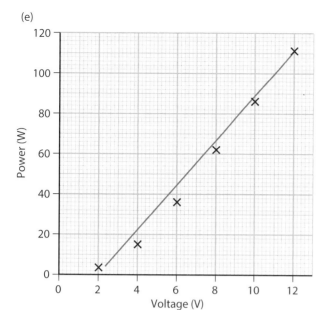

Labels and units for axes (1)
All points plotted correctly and suitable curve or line of best fit (1)
Suitable scales (1)

(f) The higher the power the brighter the bulb (1). As the voltage is increased, the power and therefore the brightness of the bulb increases (1). The relationship is not directly proportional / increase slows at higher voltage (1).

(g) (i) $0.05 \times 100 / 8.6 = 0.58$ (0.6) % (1)
(ii) $0.5 \times 100 / 10 = 5$% (1)

(h) Measurement which gives highest % error / the voltage measurement (1). Correct possible range of error 9.5–10.5 V (1).

(i) Transformation of equation to work done = time × power (1); 2580 (1)

2 (a) 8 volts – any number between 55 and 65
12 volts – any number between 135 and 145 (both needed for 1 mark), both need to have 2 decimal places.

(b) Convert time to hours and convert watts to kilowatts: 10 hours and 0.0153 kW (1)
Calculate number of kilowatt hours: 0.153
Calculate cost: 1.53 (pence) (1)
Display to two significant figures: 1.5 (pence) (1)

(c) Different bulb used (1); wires may not have cooled down between each investigation (1)

(d) At low voltage it is directly proportional (1). As the voltage increases, the increase in current is not directly proportional (1). At high voltages the increase in voltage does not cause an increase in current of the same percentage / proportion (1).

3 (a) (i) The length, diameter and material of the wire need to be controlled as they affect the resistance of the wire (1). The resistance needs to be controlled as a bigger resistance would restrict the flow of current through the wire (1).
(ii) Temperature (1); measure the current in the circuit immediately after applying voltage/closing the switch (1).

(b) Any two linked ideas (4):
Repeat with different types of bulbs at all voltages; to test if all bulbs behave in the same way.
Carry out test across bigger range of voltage; to check the pattern continues at higher voltage.
Carry out test using smaller increments in change in voltage; to see where relationship is no longer directly proportional

Section 2
(pages 60–63)

4 Your plan should include or refer to:
- a hypothesis
- equipment, techniques and /or procedures
- risks
- control variables
- dependent variable – how it will be measured, units and the precision of measurements to be taken
- independent variables – the range of measurements/ categories to be used and how they will be measured, the intervals to take measurements
- data analysis.

Example answer (12):
The hypothesis in this investigation is that plants grown in fertilisers which include nitrogen, potassium and phosphorus will grow faster than plants grown in soil with no fertiliser. This is because nitrogen is used to make amino acids for protein production. Phosphorus is needed for the formation of healthy shoots and roots and formation of DNA. Potassium is needed in the enzyme involved in respiration and photosynthesis.

Equipment:
- 20 radish seeds
- 20 small plastic pots
- soil for 20 pots (this needs to be taken from the same bag and should not include NPK fertilisers)
- distilled water
- ruler
- NPK fertiliser (made up to four concentrations; 0% fertiliser, 50% of recommended concentration on packaging, 100% of recommended concentration on packaging, 150% of recommended concentration on packaging)
- measuring cylinders
- lamp of set light intensity.

Risks:
You should wear gloves when handling the fertilisers.

Control variables
- Volume of soil should be the same in each pot.
- Light intensity should be controlled by growing all plants under the same lamp.
- Volume of water used to water the plants. Plants should be watered at the same time each day.
- All plants should be grown at the same temperature; this could be done by growing plants in a small greenhouse.
- Volume of fertiliser added to each plant should be controlled (change the concentration of fertiliser).

Dependent variable
You could measure the growth of the plants every 24 hours using a ruler. Measure to the nearest mm. Measure for 2 weeks

Method:
- Put equal volumes of soil into 20 small plastic pots.
- Plant one radish seed into each pot.
- Water with 10 cm^3 of distilled water. Divide the pots up into four groups. For group 1 add no fertiliser, for group 2 add 50% of the recommended volume of fertiliser, for group 3 add 100% of the recommended fertiliser and for group 4 add 150% of the recommended fertiliser.
- Place all plants under a lamp (and in the greenhouse).
- After 24 hours measure any growth with a ruler to the nearest mm.
- Repeat watering every 24 hours.
- Repeat measuring every 24 hours for 2 weeks.
- Calculate the average growth rate (mm/hour) for each group of plants.

5 Your answer should include some or all of the points below (8):
- It is not clear how much soil there is.
- It is not clear if temperature/light/minerals are controlled during the procedure.
- Readings are not repeated.

- It is not clear when the water is added each day.
- Measurements are not recorded to the same degree of accuracy.
- It is not clear how the height will be measured.
- It is not clear on how many days the plant will be watered.
- Non-uniform volume range.
- Data partially support conclusion.
- Insufficient data at 80 cm^3 and above.
- Result at 100 cm^3 may be an anomaly.

Unit 5: Principles and Applications of Science II

Revision test 1

Section A: Organs and systems
(pages 65–71)

1. (a) C (1)
   (b) Tachycardia is when the heart is beating faster than normal, at more than 100 bpm when resting (1).
   (c) 108 − 62 = 46 bpm (1)
       46/62 × 100 (1) = 74% (1)
   (d) Ventricular systole (1)
   (e) The young man feels 'dizzy' because not enough oxygen-rich blood (1) is being pumped to the **brain/head/body** (1).
       The young man is 'short of breath' because not enough blood is being pumped to the lungs to exchange carbon dioxide with oxygen (1), the body's response to this is to **breathe faster/ more deeply**. (1)

2. (a) A (1)
   (b) A bronchiole (1)
       B epithelial cell (of alveolus) (1)
       C capillary (1)
   (c) They have thin walls (1) shorten the diffusion pathway (1) and hollow shape (1) increases surface area for diffusion (1).
   (d) Fewer 'working alveoli', means a reduced surface area in the lungs. (1) This means that less oxygen can diffuse into the blood/carbon dioxide out of the blood (1). Not as much oxygen available for respiration in muscles, so less energy (slower/tired more easily) (1).
   (e) Normal cardiac output: 70 × 62 (1) = 4340 ml/min (1)
       Woman marathon runner stroke volume:
       4590/51 (1) = 90 ml (1)
   (f) D (1)
   (g) The female marathon runner's body seems to have adapted to provide more oxygen into her body by increasing her rate of breathing from 12 **to 18 breaths per minute** (1). **Her heart also seems to have adapted by having a greater stroke volume (90 compared to 70 ml), which allows a greater volume of blood to be pumped at any one time** (1). **Her heart is also stronger, as she has a low resting heart rate (51 compared to 62 bpm)** (1).

3. (a) Experimental error/an item of equipment used to sample or test the blood were contaminated with some glucose (1).
   (b) The concentration of glucose is very high compared to patients A and B (1).
   (c) Proteins usually remain in the blood because the molecules are too **big** (1) **to pass through the Bowman's capsule into the filtrate** (1). **If there are proteins in the urine, it indicates they have passed through and there could be some damage to the kidney** (1).
   (d) Example answer (6):
       - All small molecules in the blood are filtered out of the blood in the Bowman's capsule.
       - Those that are essential or needed in differing amounts are reabsorbed.
       - All glucose and, depending on the body's needs, some water and some mineral ions are reabsorbed.
       - Urea is a waste product and is excreted/not reabsorbed.
       - Water is reabsorbed in the descending (first section of the) loop of Henle. It moves by osmosis from the filtrate back into the blood because the blood is more concentrated.
       - As more water is reabsorbed the concentration of the filtrate increases, and some ions and glucose molecules diffuse out of the loop of Henle (ascending) into the blood.
       - At the end of the loop of Henle the concentrations of the filtrate and blood are about the same, and any remaining glucose is moved out by active transport.

Section B: Properties and uses of substances
(pages 72–78)

1. (a) (i) C_nH_{2n+2} (1)
       (ii) Alkenes (1)
   (b) (i) Contains a carbon-to-carbon double bond (1)
       (ii) With a saturated hydrocarbon the bromine water stays **orange** (1).
            With an unsaturated hydrocarbon the bromine water turns **colourless** (1).
   (c) **Step 2** (propagation): $C_3H_8 + Cl\bullet \rightarrow \bullet C_3H_7 + HCl$ (1)
       $\bullet C_3H_7 + Cl_2 \rightarrow C_3H_7Cl + Cl\bullet$ (1)
       Step 3 (termination): Any one of:
       $Cl\bullet + Cl\bullet \rightarrow Cl_2$
       $\bullet C_3H_7 + Cl\bullet \rightarrow C_3H_7Cl$
       $\bullet C_3H_7 + \bullet C_3H_7 \rightarrow C_6H_{14}$ (1)
   (d)

   Arrow from double bond to bromine molecule (1)
   Arrow from Br–Br bond to bromine atom of bromine molecule (1)
   Either carbocation intermediate (1)
   Arrow from lone pair of Br$^-$ ion to positively charged carbon atom of the carbocation (1)

2. (a) (i) $3NaOH + \mathbf{Al(OH)_3} \rightarrow Na_3Al(OH)_6$
            Correct formulae (1); correct balancing (1)
       (ii) Sodium hydroxide is a **base which can only react with acids** (1).
            Iron(III) oxide and titanium oxide are both basic oxides (1), **whereas aluminium hydroxide is amphoteric and therefore can act as an acid** (1).

(b) (i) Dilute sulfuric acid is added slowly to the filtrate (until precipitation stops) (1).
(ii) $2Al(OH)_3 \rightarrow Al_2O_3 + 3H_2O$ (1)
(c) (i) Titanium/magnesium reacts with oxygen/nitrogen in air (1), but not with argon (1).
(ii) The oxidation number of titanium changes from +4 (in $TiCl_4$) to 0 (in Ti) (1) and so it has been reduced (1). The oxidation number of magnesium changes from 0 (in Mg) to +2 (in $MgCl_2$) (1) and so it has been oxidised (1).

3 (a) $Pb(s) + ½O_2(g) \rightarrow PbO(s)$
Correct equation (1); correct state symbols (1)
(b) The enthalpy change when one mole of a substance is formed from its elements (1) in their standard states (1) measured at a stated temperature (usually 298 K) and at a pressure of 100 kPa (1).
[Note: the substance formed does **not** have to be in its standard state. The standard enthalpy change of formation of liquid water is -286 kJ mol^{-1}, whereas the standard enthalpy change of formation of gaseous water is -242 kJ mol^{-1}.]
(c) $\Sigma \Delta_f H°(\text{reactants}) = -\mathbf{(2 \times -219) + \Delta_f H°[PbO_2(s)]}$ (1)
$\Sigma \Delta_f H°(\text{products}) = \mathbf{-735}$
$\Delta_r H° = \Sigma \Delta_f H°(\text{products}) - \Sigma \Delta_f H°(\text{reactants})$
$\mathbf{-10 = -735 - \{(2 \times -219) + \Delta_f H°[PbO_2(s)]\}}$ (1)
$\Delta_f H°[PbO_2(s)] = \mathbf{-735 - (2 \times -219) - (-10)}$ (1)
$= \mathbf{-287 \text{ kJ mol}^{-1}}$ (1)

4 **Example answer** (6):
- The mercury cell is expensive to construct, whereas the diaphragm and membrane cells are relatively cheap.
- Mercury is toxic and has to be removed from the waste material before it can be discharged.
- The diaphragm has to be regularly replaced and this increases the maintenance costs.
- The membrane cell has low maintenance costs.
- The mercury cell produces high-purity sodium hydroxide at the concentration required.
- The diaphragm cell produces low-purity sodium hydroxide at a low concentration. This solution needs further evaporation and this increases the running costs.
- The membrane cell produces high-purity sodium hydroxide but of a lower concentration than the mercury cell.
- The membrane cell uses the least amount of electricity.
- Reach a justified conclusion as to which method is the best.

Section C: Thermal physics, materials and fluids
(pages 79–85)

1 (a) (i) No work has been done because $W = p\Delta V$ (1) and there has been no change in **volume** (1).
(ii) ΔU is **negative** (1) as $W = 0$ and Q is **negative** (1), because **heat was transferred out of the system** (1).
(b) The heat capacity of the water is:
$2.40 \times 10^{-3} \text{ m}^3 \times 1000 \text{ kg m}^{-3} \times 4.18 \text{ kJ kg}^{-1} \text{ K}^{-1}$
$= 10.03 \text{ kJ K}^{-1}$ (1).
So the total heat capacity of the calorimeter, water bath and water
$= \mathbf{10.03} + 1.25 \text{ kJ K}^{-1} = \mathbf{11.28} \text{ kJ K}^{-1}$ (1)
$\Delta U = Q = \mathbf{-11.28} \text{ kJ K}^{-1} \times (23.1 °C - 20.2 °C) = \mathbf{32.7 \text{ kJ}}$ (1)
(sign and unit essential)
(c) There would be a smaller temperature rise (1) because atmospheric pressure would be lower/because the gases could expand (1) and the gases would do positive work W, meaning a smaller value for Q (1).

2 (a) $-78 + 273.15 = 195.15 \text{ K}$ (1)
(b) The supply of energy for the latent heat of vaporisation comes from the sensible heat rejected as **the temperature of the liquid falls** (1).
This is followed by the latent heat of fusion given out as the liquid then freezes to form dry ice (1).

(c) ΔU for the condensation of gas to liquid is **negative**, because **latent heat is being given out** (1). W is negative because **compression work is being done on the gas** (1) – **its volume change is negative. So, adding those two together, Q must be negative** (1) **and quite large, as it is the sum of the other two negative quantities. That means a lot of heat has to be removed.**

3 (a) Polyethylene (1)
(b) Stainless steel (1), because it has the highest Young's modulus (1).
(c) Concrete and glass are used in situations where they are under compressive stress (1) because they are known to fracture under quite small tensile stress (1).
(d) (i) Engineers should use **yield strength** (1) because body panels should not be permanently **deformed** by normal stresses (1).
(ii) ABS plastics = $40 / 1050 \text{ MPa m}^3 \text{ kg}^{-1} = \mathbf{0.038}$ (1)
Aluminium = $95 / 2700 \text{ MPa m}^3 \text{ kg}^{-1} = \mathbf{0.035}$ (1)
Stainless steel = $502 / 7870 \text{ MPa m}^3 \text{ kg}^{-1} = \mathbf{0.064}$ (1)
So yield strength is highest for stainless steel (nearly twice the others) (1).

4 (a) When forces acting on the surface are parallel to the surface they are acting on (1).
(b) Viscosity remains constant at a given temperature (1) because of the linear increase in stress as shear rate increases (1).
Molecules are not oriented by flow (1) because they are isotropic (1).
(c) **Example answer** (6):
Interpretation of the graphs:
- The effective viscosity of a fluid is the ratio (shear stress)/(rate of shear strain).
- Straight line through the origin means constant viscosity.
- Increasing slope means increasing viscosity and vice versa.
- Lines that pass through the origin show some flow at the lowest stress levels – i.e. liquid behaviour.
- Intersecting the stress axis above the origin means initial solid behaviour, with a threshold stress to cause flow.

Non-drip paint:
- Comparison: High initial viscosity that falls when brushed.
- Explanation: Ease of application. Rapid setting to high-viscosity film that doesn't drip or run.

Motor oil:
- Need to maintain a steady viscosity so that the lubricating films inside the engine's bearings and other moving parts remain thick enough to keep the surfaces apart, but do not create too large a **viscous drag** that would slow the engine down and waste energy.
- Comparison: Viscosity independent of shear rate.
- Explanation: Maintains lubricating film at all speeds. Avoids unduly large viscous drag forces.

Body armour gel:
- Need to slow or stop a bullet or a knife.
- Comparison: Low viscosity with rapid thickening.
- Explanation: Flexible to allow wearer to move gently. Very high viscosity (almost solid) when struck or pierced quickly.

Moist clay:
- The line crosses the stress axis at quite a high value for zero rate of shear strain.
- Comparison: Threshold stress before flow, which then develops rapidly.
- Explanation: Construction engineers must design so as not to exceed the threshold stress. Exceeding it causes subsidence or landslips due to rapid onset of fluid behaviour.

Revision test 2

Section A: Organs and systems
(pages 86–92)

1. (a) It is passive because energy is not required for it to happen (1).
   (b) There is a higher concentration in the alveoli than in the blood (1), and diffusion occurs down a concentration gradient from a high to a low concentration, so the carbon monoxide moves from the alveoli into the blood (1).
   (c) Three from the following points (3):
   - permeable membranes (pores large enough for molecules to diffuse through)
   - substances diffusing are soluble
   - layer of moisture within the alveoli
   - substances diffusing or the pores are not charged.

   (d) Osmosis refers only to the movement of water molecules (1).
   (e) ADH is responsible for opening the channels in the collecting duct (1).
   Makes the collecting duct more permeable and easier for water molecules to move from the filtrate into the medulla (1).

2. (a) Two of the following (2):
   - Not ethical to use humans (*Daphnia* has a simple nervous system – feels less pain).
   - Abundant and easy to obtain.
   - Transparent so can see heart.

   (b) Place *Daphnia* in solutions of nicotine (1); place onto well slide and immobilise using cotton wool strands (1); view through microscope (1); use dots, clicker, calculator key to count bpm (1).
   (c) Three of the following: temperature, age, size, sex, aspects of pretreatment (3).
   (d) Students create serial solutions, i.e. solutions containing 0, 1%, 10% and 100% nicotine solution (1).
   Test which solutions produce the most measurable results where an effect can be seen without killing the *Daphnia*) (1).

3. (a) B (1)
   (b) Botokil prevents the carrier proteins in the fungal cells having energy (1), so they will not be able to move substances across the cell membrane against a concentration gradient, so the fungal cells will die (1).
   (c) Made up of two layers of phospholipids (1); phospholipid made up of a phosphate group, which is hydrophilic (water-loving) and is on the outside of the bilayer (1) and the lipid group, which is a hydrophobic tail (water-hating) and is always on the inside of the bilayer (1).

4. (a) D (1)
   (b) Four of the points below (4):
   - Both arteries and veins lined with smooth endothelium.
   - Arteries – thick wall, smooth muscle, narrow lumen, no valves
   - Veins – thin wall, very little muscle, wide lumen, valves.

   (c) Sex (1), gender (1), genetics (1)
   (d) Example answer (6):
   - Improve diet – more vegetables, protein, less fat, less carbohydrate.
   - Improved diet will control weight which will reduce high blood pressure.
   - Improved diet may reduce cholesterol levels.
   - Do more exercise.
   - More exercise will make heart stronger; lower blood pressure; helps reduce/control weight.
   - Stop smoking.
   - Stopping smoking will mean the heart doesn't have to work as hard (nicotine stimulates body to produce adrenaline, which in turn makes the heart beat faster; it also means that less carbon monoxide/more oxygen is absorbed into the blood, so heart has to work less hard to gain oxygen) and reduces damage to arteries (aneurysm), leading to fatty deposits (atheroma).

Section B: Properties and uses of substances
(pages 93–98)

1. (a) (i) $C_{18}H_{38}$ (1)
   (ii) Any two from (2):
   - The compounds have similar chemical properties.
   - The compounds have graded physical properties.
   - Each compound differs from the previous one in the series by a $-CH_2-$ unit.

   (b) (i) Cracking produces shorter chain hydrocarbons (1) because these are in greater demand than the long-chain hydrocarbons (1).
   Cracking produces alkenes (1) and these are used to make polymers (1).
   (ii) $C_{16}H_{34} \rightarrow C_8H_{18} + 2C_4H_8$ or $C_{16}H_{34} \rightarrow C_8H_{18} + 4C_2H_4$ (2)
   Correct formula for hexadecane (1)
   Correct cracking equation (1)

   (c) (i) A π bond is formed by the sideways overlap of two p orbitals (1).

   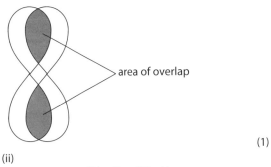

   (1)

   (ii)

   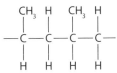

   Correct structure (1)
   Extension bonds shown (1)

2. (a) $C_3H_8 + 5O_2 \rightarrow 3CO_2 + 4H_2O$
   All formulae correct. (1)
   Equation correctly balanced. (1)
   (b) The enthalpy change measured at a stated temperature (usually 298 K) and 100 kPa pressure (1) when one mole of a substance is burned (1) completely in oxygen (1).
   (c) (i) Heat energy transferred = $200 \, g \times 4.18 \, J\,g^{-1}\,K^{-1} \times 50.3 \, K$
   $= 42\,050.8 \, J$ (1)
   $= 42.051 \, kJ$ (1)
   $= 42.1 \, kJ$ (to 3 s. f.) (1)
   (ii) Some energy will be lost to the surroundings (1).

   (d)

   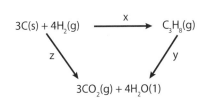

   $x + y = z$
   or $x = z - y$
   or $x = (3 - 394) + (4 - 286) - (-2219)$ (1)
   $\Delta_f H$ [propane] $= -107 \, kJ\,mol^{-1}$ (1)

3 (a) $2H_2O(l) + 2e^- \rightarrow H_2(g) + 2OH^-(aq)$
       correct reactants (1)
       correct products (1)
   (b) Sodium hydroxide/NaOH (1) can be used for laboratory reactants/during food processing/during paper manufacture (1).
   (c) Electrolysis requires a very large amount of electricity (1) therefore costing a lot of money/potentially causing pollution (from burning of fossil fuels) (1).
       or
       Older plants may use mercury cells (1) therefore leading to potential waste toxicity (1).
   (d) The reaction is exothermic (1) because the enthalpy value is negative (1) which means that heat energy is given out (1).

4 **Example answer** (6):
   - Both heterogeneous catalysts.
   Haber process:
   - Nitrogen and hydrogen molecules adsorb onto the surface of the iron.
   - The covalent bonds within the molecules are weakened.
   - The weakening of the bond lowers the activation energy for the reaction.
   - Reaction between nitrogen and hydrogen takes place at the surface of the iron.
   - After reaction the ammonia molecules desorb from the surface of the iron.
   Contact process:
   - Sulfur dioxide reacts with the vanadium(V) oxide catalyst.
   - $V_2O_5(s) + SO_2(g) \rightarrow V_2O_4(s) + SO_3(g)$
   - The vanadium(V) oxide catalyst is regenerated by reaction with oxygen.
   - $V_2O_4(s) + \frac{1}{2}O_2(g) \rightarrow V_2O_5(s)$
   - Both reactions have a lower energy than the direct reaction between $SO_2(g)$ and $O_2(g)$.

Section C: Thermal physics, materials and fluids
(pages 99–107)

1 (a) (i) Stress is proportional to strain (1).
       (ii) Work hardening (1)
       (iii) B (1)
       (iv) CD (Accept BD) (1)
   (b) (i) $F = k\Delta x$ (1)
       $\Delta x = F/k = (4000/25000)$ m (1)
       $= 0.16$ m (1)
   (c) $W = \frac{1}{2} k\Delta x^2$ or $W = \frac{1}{2} F\Delta x$
       $= \frac{1}{2} \times 4000 \times 0.16$ or $\frac{1}{2} \times 25000 \times 0.16^2$ (1)
       $= 320$ J
   (d) Any two differences, such as:
       Creep – gradual failure, more likely at higher operating temperatures
       Fatigue – cyclic loading/repeated loading and unloading, fracture occurs suddenly (2)

2 (a) (i) $W = mgh$
       or
       Work done = mass × g × height (1)
       (ii) $W = 2270 \times 9.81 \times (1.2 - 0.3)$ (1)
       $W = 20041$ J (1)
       (Accept 20 430 J if $g = 10$ N/kg has been used.)
   (b) (i) ΔV = change in volume (1)
       (ii) $p = W/\Delta V$ (1)
       $= 20000/800 \times 10^6 = 25$ MPa (1)
       (iii) Efficiency will be less than 100% (1) due to friction, air resistance, heat losses, etc. (1)
   (c) Efficiency = 20 000/45 200 (1)
       = 44% (1)

3 (a) Spring: stores energy, exchanging it between kinetic energy (motion) and potential energy (position) (1).
       Damper: transforms kinetic energy into heat (1).
   (b) Any two of: viscosity; specific heat capacity; freezing and boiling points, fire resistance (2).
   (c) Any two of: Young's modulus; yield strength; fatigue limit (2).
   (d) First Law: no heat is initially transferred, but work is done on the fluid and that increases its internal energy, hence the rise in temperature (1).
       Second Law: the transfer of work into thermal energy is irreversible – that energy could never be fully recovered as work (1).

4 (a) Heat energy absorbed or given out when a solid melts or a liquid solidifies (1)
   (b) $Q = Pt = 1000$ W $\times 300$ s $= 300$ kJ (1)
       $Q = mL + mc\Delta T$
       $1000 \times 300 = 0.8 \times 330\,000 + 0.8 \times 4200 \times \Delta T$ (1)
       $0.8 \times 4200 \times \Delta T = 1000 \times 300 - 0.8 \times 330\,000$ (1)
       $\Delta T = 11\,°C$ (1)
   (c) $COP = T_c/(T_h - T_c)$ (1)
       $= 255/40 = 6.375$ (1)
   (d) The coil material should have high strength, high ductility, high thermal conductivity and low cost. Density is relevant to costing, because the prices are given in £ per tonne. So first calculate costs per m³:
       aluminium alloy: $1.26 \times 2.7 = 3.4$ £/m³
       copper: $5.0 \times 8.9 = 44.5$ £/m³
       brass: $2.5 \times 8.7 = 21.8$ £/m³
       Strength – the coils need to be strong in order to resist the pressure from the fluid inside. Brass has the highest strength of the three shown.
       Ductility – a high percentage elongation indicates the material can be readily drawn into tubes and bent into coils.
       Thermal conductivity – high thermal conductivity means that the transfer of heat from the freezer compartment to the refrigerant is more efficient. Thinner walled tubes, made possible by high ductility also improve the heat transfer rate for a given surface area of tubing.
       Copper has the optimum thermal conductivity.
       Cost – aluminium is the cheapest material. It also has a weight advantage when used in freezers on vehicles or aircraft.
       Judgement based on the evidence: for this application, thermal conductivity is probably the most important, therefore copper could be chosen. However, aluminium has greater strength, adequate ductility and lower cost, so would be a reasonable alternative. Brass, while almost as strong and ductile as copper and half the price, has such a poor thermal conductivity that overall it does not align well with the requirements.

Unit 7: Contemporary Issues in Science

Revision task 1
(Pages 126–143)

1 Continuation of answer regarding identification of the scientific issue may include:
   - malaria affects around 219 million people a year and causes around 600,000 deaths; in Africa 10 new cases of malaria every second and a child dies every 2 minutes
   - malaria is a disease caused by the blood parasite *Plasmodium*, which is transmitted by mosquitoes

- social problems: families bereaved when children or parents die; ill parents cannot work and support their children; children lose time in school; any other sensible point
- economic problems: for individual families when parents cannot work or die from malaria; high medical costs for families and state; economic cost to countries as work force not very productive due to malarial illness; any other sensible points

Continuation of answer regarding treatment of malaria may include:
- use of insecticide sprays in the environment and the home
- environmental problems from most effective insecticide (DDT) – enters food chain and kills top carnivores
- spraying insecticide regularly very expensive
- antimalarial drugs – not always effective and expensive
- malarial parasite becoming resistant to drugs
- no really effective vaccines available
- vaccine for small children – but little health infrastructure so hard to get people to bring babies for vaccines regularly
- simple insecticide-impregnated nets very effective but have to persuade people to use them
- any other sensible points.

Continuation of answer regarding downsides of malaria treatment may include:
- cost for poor communities
- lack of effective vaccines
- cost of vaccines
- problems of getting people to have vaccines or take medicines regularly
- use of artemisinin – rapid action, very effective but resistance appearing
- artemisinin combined therapies ACTs – much more effective but now resistance emerging to the combination drug
- other sensible points.

Continuation of answer relating treatment of malaria to the articles may include:
- Reference Article 3 for a series of comments on the problems of using antimalarial drugs, including use of genome analysis.
- Highlight the costs and the problems of getting hi-tech equipment to where it is needed.
- Any other sensible points.

Continuation of answer regarding drug resistance may include:
- Resistance often emerges in Greater Mekong Subregion in South East Asia.
- Use Article 3 to give reasons why this might happen.

Continuation of answer regarding tackling drug resistance may include:
- Role of artemisinin-based combination therapy (ACT) in global malaria control.
- ACT provides an inexpensive, short-course treatment that would also help protect against the development of drug resistance.
- Artemisinin is a very fast-acting drug which means that within 12 hours of starting treatment around half of the parasites in the body are removed.
- It is combined with another drug that usually works more slowly, killing the remaining malaria parasites.

Continuation of answer regarding global implications of drug-resistant malaria may include:
- highlight individual cost
- cost to the country
- cost of developing new treatments
- global warming means temperature at surface of Earth increasing and is affecting rainfall
- mosquitoes which carry malaria can only survive and breed successfully in warm, wet conditions
- more countries – for example in Europe around the Mediterranean coast – will have these conditions and so the mosquitoes will be able to spread, bringing malaria with them
- any other sensible points.

2. Continuation of answer regarding institutions may include:
   - WHO is a specialised part of the United Nations and is very influential. It collects data on diseases from all over the world and issues guidelines for treatments, flags up epidemics and pandemics etc. It put together the 2012 World Malaria Report based on data from 104 countries with endemic malaria. Very influential globally (WHO publications website, Wikipedia).
   - The Institut Pasteur is in Paris. It is an internationally renowned centre for research into microbiology and disease – based on the work on microbiology, vaccines etc. of Louis Pasteur.

   Continuation of answer regarding people may include:
   - Sir Ronald Ross (1902) – a doctor who demonstrated that malaria is spread by mosquitoes (Wikipedia, Nobel Prize website).
   - Charles Louis Alphonse Laveran (1907) discovered the role of parasites in malaria.
   - Julius Wagner-Jauregg (1927) – doctor who discovered a treatment for paralysis from syphilis by infecting patients with malaria to give them a high fever which cured the paralysis (Nobel Prize website, Wikipedia).
   - Paul Hermann Müller (1948) discovered DDT as insecticide against several arthropods including mosquitoes.
   - Professor David Conway is very active in antimalaria research. He is a professor at the London School of Hygiene and Tropical Medicine, works in the UK and African countries, has published over 170 research articles, and is well known in his field (LSHTM website).
   - Any other scientists you notice in the articles and can reference.

3. Continuation of answer may include:
   - Worldwide Antimalarial Resistance Network: WHO.
   - *Nature Genetics* – highly reputable peer-reviewed journal
   - Wellcome Trust Genome Campus has produced at least one Nobel Prize winner - scientists who work there are top calibre and liaise with other scientists around the world.
   - References a study with 1000 patients – this is a relatively large sample and therefore the results are likely to be reliable and valid.

4. Continuation of answer on genome analysis may include:
   - Genome analysis reveals the DNA sequences of the malarial parasites.
   - We can use this information to show us when the malarial parasite begins to develop resistance to a drug and change the treatments used in this area.
   - Use the information to help develop potential medicines and vaccines.
   - Any other sensible points.

   Continuation of answer on development of new insecticides may include:
   - Discuss the need for new insecticides and the criteria for their development – safe for people and the environment.
   - Make any other points about potential new research and development you have.

5. As well as the suggestions given in the guided answer, you could also:
   - Explain why malaria is becoming even more dangerous as parasites become resistant to more of the antimalarial drugs.
   - Describe how we have started to overcome resistance through genomics etc.
   - Raise the possibility of super-parasites resistant to new drugs as well.
   - Explain carefully how mosquito nets help prevent mosquitoes transmitting disease.
   - Go on to explain how insecticide-impregnated nets give better protection than nets alone.
   - Discuss the pros and cons of them – including that they are cheap and easy for people to use.

- Produce some supporting data for example a bar chart using data from Article 3 on how quickly the malaria parasite has developed resistance to new drugs.
- Write a bit more about how genome analysis can help scientists understand how the malaria parasite develops immunity to drugs and how they can use it to develop new drugs.

Revision task 2

(pages 153–161)

1

> The science issue identified in these articles is based around the need for alternative fuel sources to generate electricity. These bullet points give you an idea of the content you need to cover. You must make a plan and write in complete sentences. So for example, you need to cover the difference between nuclear fission and nuclear fusion. They are discussed in both Article 2 and Article 3. You only need to talk about it once, referencing both articles – then you can go on to discuss the differences and similarities between the ITER project using a tokamak and the stellerator project.

Article 1:
- light poverty – lack of electricity for 1.1 billion people worldwide
- alternative fuels used to give light, for example kerosene, fires can cause death through fires and lung disease
- problems of delivering conventional electricity supplies in developing nations
- solar-powered LEDs as a solution
- any other points.

Article 2:
- the challenge of making electricity using nuclear fusion rather than fission or fossil fuels – safety, economic, timescale
- what nuclear fusion involves
- what would be the advantages over fission
- the issues of developing a tokamak
- the ITER project – practical issues with producing nuclear fusion, cost, international cooperation
- any other sensible points.

Article 3:
- stellarators as an alternative way of achieving nuclear fusion for electricity generation
- advantages of fusion over fission – safety angle
- problems of developing fusion generator
- comparison of tokamak and stellarator
- problems of cost and rates of progress.

2

> You need to write about at least six of the following organisations and people – at least one from each article. You should have looked them up and made suitable notes before you make your plan and then answer in complete sentences.

- United Nations Environment Programme (UNEP)
- Eric Rondolat, Chief Executive Officer, Philips Lighting
- Global Off-Grid Lighting Association
- World Bank
- ITER.org
- Culham Centre for Fusion Energy – ccfe.ac.uk
- Max Planck Institute of Plasma Physics (IPP)
- Wendelstein 7-X stellarator nuclear fusion device
- Professor Thomas Klinger, scientific director for the stellarator project
- Professor Tony Donné, programme manager at EUROfusion, coordinates JET and ITER.

3

> Take note which article(s) you are asked about. In this case the question covers Article 3. Make suitable notes before you make your plan. Remember you will get marks for a clear coherent discussion.

Some of the points covered should include:
- an explanation of what is meant by a valid judgement and reliable data
- any immediate issues or bias in any of the articles
- any points which redress that potential bias
- Article 2 – an essay written by a university physics student, so it could have weaknesses, lack of understanding if student hasn't fully understood the concepts **but** student has based essay on reliable sources and when the references are checked, the essay content appears sound
- student references ITER, widely renowned organisation with scientists from 35 nations collaborating, so unlikely to be lying/bending facts as many reputations on the line
- quotes Culham Centre for Fusion Energy which is the best functioning current tokamak
- everything about tokamaks in Article 3 confirms what is said in Article 2, which increases likelihood of validity
- quotes from highly reputable scientists including people working on tokamaks (Professor Tony Donné who coordinates JET and ITER) and stellarators (Professor Thomas Klinger, scientific director for the stellarator project)
- any other sensible points.

4
- Discuss possible ways of powering safe, clean electric light sources for homes in the developing world – give some data to explain why it is necessary and at least two possible ways of generating the electricity needed to light homes on a small scale. Relate to Article 1.
- Further research and development into the use of nuclear fusion as a potential source of the energy needed to generate electricity on a large scale – relate to Articles 2 and 3.

5

> Make suitable notes before you make your plan. Make sure you keep your target audience in mind throughout. Use information from the three articles provided for this task.

Points to cover include:
- Living in the UK it is almost impossible to imagine life without electric lighting.
- Most of the electricity we use to light our homes is generated by burning fossil fuels. This produces carbon dioxide, which is adding to the greenhouse effect, linked to global warming and climate change.
- Fossil fuels will get more and more expensive as they start to run out – they are non-renewable.
- Around 1.1 billion people have no electric lighting in their homes at all.
- Often use kerosene and other fuels which risk fires starting and damage health – can cause death.
- Need safe, environmentally friendly and cheap ways to produce electricity.
- One way of tackling the problems – large-scale, for example developing ways of producing electricity from nuclear fusion.
- Describe simply how the processes might work, big advantages and some of the problems.
- Another way – small-scale – individual or village-level solutions especially useful in developing world. For example solar lighting, investigate other possibilities.
- Include lots of references to articles given and further reading.
- Any other sensible points.

Notes

Notes

Notes